# Selberg Trace Formulae and Equidistribution Theorems for Closed Geodesics and Laplace Eigenfunctions: Finite Area Surfaces

# Recent Titles in This Series

464 **John Fay,** Kernel functions, analytic torsion, and moduli spaces

463 **Bruce Reznick,** Sums of even powers of real linear forms

462 **Toshiyuki Kobayashi,** Singular unitary representations and discrete series for indefinite Stiefel manifolds $U(p,q;\mathbb{F})/U(p-m,q;\mathbb{F})$

461 **Andrew Kustin and Bernd Ulrich,** A family of complexes associated to an almost alternating map, with application to residual intersections

460 **Victor Reiner,** Quotients of coxeter complexes and $P$-partitions

459 **Jonathan Arazy and Yaakov Friedman,** Contractive projections in $C_p$

458 **Charles A. Akemann and Joel Anderson,** Lyapunov theorems for operator algebras

457 **Norihiko Minami,** Multiplicative homology operations and transfer

456 **Michał Misiurewicz and Zbigniew Nitecki,** Combinatorial patterns for maps of the interval

455 **Mark G. Davidson, Thomas J. Enright and Ronald J. Stanke,** Differential operators and highest weight representations

454 **Donald A. Dawson and Edwin A. Perkins,** Historical processes

453 **Alfred S. Cavaretta, Wolfgang Dahmen, and Charles A. Micchelli,** Stationary subdivision

452 **Brian S. Thomson,** Derivates of interval functions

451 **Rolf Schön,** Effective algebraic topology

450 **Ernst Dieterich,** Solution of a non-domestic tame classification problem from integral representation theory of finite groups ($\Lambda = RC_3, v(3) = 4$)

449 **Michael Slack,** A classification theorem for homotopy commutative H-spaces with finitely generated mod 2 cohomology rings

448 **Norman Levenberg and Hiroshi Yamaguchi,** The metric induced by the Robin function

447 **Joseph Zaks,** No nine neighborly tetrahedra exist

446 **Gary R. Lawlor,** A sufficient criterion for a cone to be area-minimizing

445 **S. Argyros, M. Lambrou, and W. E. Longstaff,** Atomic Boolean subspace lattices and applications to the theory of bases

444 **Haruo Tsukada,** String path integral realization of vertex operator algebras

443 **D. J. Benson and F. R. Cohen,** Mapping class groups of low genus and their cohomology

442 **Rodolfo H. Torres,** Boundedness results for operators with singular kernels on distribution spaces

441 **Gary M. Seitz,** Maximal subgroups of exceptional algebraic groups

440 **Bjorn Jawerth and Mario Milman,** Extrapolation theory with applications

439 **Brian Parshall and Jian-pan Wang,** Quantum linear groups

438 **Angelo Felice Lopez,** Noether-Lefschetz theory and the Picard group of projective surfaces

437 **Dennis A. Hejhal,** Regular $b$-groups, degenerating Riemann surfaces, and spectral theory

436 **J. E. Marsden, R. Montgomery, and T. Ratiu,** Reduction, symmetry, and phase mechanics

435 **Aloys Krieg,** Hecke algebras

434 **François Treves,** Homotopy formulas in the tangential Cauchy-Riemann complex

433 **Boris Youssin,** Newton polyhedra without coordinates Newton polyhedra of ideals

432 **M. W. Liebeck, C. E. Praeger, and J. Saxl,** The maximal factorizations of the finite simple groups and their automorphism groups

431 **Thomas G. Goodwillie,** A multiple disjunction lemma for smooth concordance embeddings

(*Continued in the back of this publication*)

Number 465

# Selberg Trace Formulae and Equidistribution Theorems for Closed Geodesics and Laplace Eigenfunctions: Finite Area Surfaces

Steven Zelditch

March 1992 • Volume 96 • Number 465 (third of 4 numbers) • ISSN 0065-9266

**American Mathematical Society**
Providence, Rhode Island

1991 *Mathematics Subject Classification.*
Primary 11F, 58F; Secondary 58G.

---

**Library of Congress Cataloging-in-Publication Data**

Zelditch, Steven, 1953–
Selberg trace formulae, and equidistribution theorems for closed geodesics and Laplace eigenfunctions: finite area surfaces/Steven Zelditch.
p. cm. – (Memoirs of the American Mathematical Society, ISSN 0065-9266; no. 465)
"March 1992."
"Volume 96 number 465."
Includes bibliographical references.
ISBN 0-8218-2526-7
1. Curves on surfaces. 2. Geodesics (Mathematics) 3. Cusp forms (Mathematics) 4. Eisenstein series. I. Title. II. Series.
QA3.A57 no. 465
[QA643]
510 s–dc20 91-44875
[516.3′6] CIP

---

**SUBSCRIPTION INFORMATION.** The 1992 subscription begins with Number 459 and consists of six mailings, each containing one or more numbers. Subscription prices for 1992 are $292 list, $234 institutional member. A late charge of 10% of the subscription price will be imposed on orders received from nonmembers after January 1 of the subscription year. Subscribers outside the United States and India must pay a postage surcharge of $25; subscribers in India must pay a postage surcharge of $43. Expedited delivery to destinations in North America $30; elsewhere $82. Each number may be ordered separately; *please specify number* when ordering an individual number. For prices and titles of recently released numbers, see the New Publications sections of the NOTICES of the American Mathematical Society.

**BACK NUMBER INFORMATION.** For back issues see the AMS Catalogue of Publications.

Subscriptions and orders for publications of the American Mathematical Society should be addressed to American Mathematical Society, Box 1571, Annex Station, Providence, RI 02901-1571. *All orders must be accompanied by payment.* Other correspondence should be addressed to Box 6248, Providence, RI 02940-6248.



**Memoirs** of the American Mathematical Society (ISSN 0065-9266) is published bimonthly (each volume consisting usually of more than one number) by the American Mathematical Society at 201 Charles Street, Providence, Rhode Island 02904-2213. Second Class postage paid at Providence, Rhode Island 02940-6248. Postmaster: Send address changes to Memoirs of the American Mathematical Society, American Mathematical Society, Box 6248, Providence, RI 02940-6248.


Printed in the United States of America.
The paper used in this book is acid-free and falls within the guidelines established to ensure permanence and durability. ∞

10 9 8 7 6 5 4 3 2 1 97 96 95 94 93 92

<u>Table of Contents</u>

## Abstract

We use harmonic analysis on a finite area hyperbolic surface $\Gamma\backslash h$ to optimally sharpen the Bowen equidistribution theory of closed geodesics. This theory is concerned with the series $\Phi(\sigma,T) = \sum_{L_\gamma \le T} \int_\gamma \sigma$, where $\sigma$ is an automorphic form on $\Gamma\backslash PSL_2(\mathbb{R})$, and where the sum runs over all closed geodesics $\gamma$ of length $L_\gamma$ less than T. Using a generalization of the Selberg trace formula, we obtain an asymptotic expansion for each $\Phi(\sigma,T)$, which reverts to the standard prime geodesic theorem if $\sigma = 1$. The principal term vanishes if $\sigma$ is orthogonal to 1, giving a new proof that closed geodesics are uniformly distributed with respect to Haar measure, together with an exponential rate of equidistribution.

One of the main steps in deriving these asymptotic expansions is to estimate certain dual spectral sums $(M+N)(\sigma,T)$, which revert to the usual winding number of the scattering phase plus eigenvalue counting function if $\sigma = 1$. When $\sigma$ is an Eisenstein series and $\Gamma = PSL_2(\mathbb{Z})$, our estimate gives a weak (signed and averaged) version of the Lindelöf hypothesis for Rankin-Selberg zeta functions.

This paper is a continuation of our earlier work on compact hyperbolic surfaces (cf. [Z.2]).

(1) Subject Classification: Primary 11F, 58F

Secondary 58G

(2) Key words: finite area hyperbolic surface, closed geodesic, cusp form, Eisenstein series, Rankin-Selberg zeta function.

Received by editor March 27, 1990 and in revised form March 4, 1991.

## § 0. Introduction

We will be concerned in this paper with a pair of closely related problems of asymptotic analysis on a (non-compact) finite area hyperbolic surface $\Gamma\backslash h$. Here, h is the hyperbolic plane and $\Gamma$ is a cofinite discrete subgroup of $G = PSL_2(\mathbb{R})$. The compact case was studied in [Z.2], of which this paper is a continuation.

The first problem is to determine the precise asymptotic distribution of closed geodesics in the unit tangent bundle. For general algebraic Anosov flows on compact quotients, it was proved by Bowen in 1972 [B] that closed orbits tend on average to become uniformly distributed with respect to Haar (or, Liouville) measure as the period tends to infinity. Bowen's theorem has since been generalized and refined by Parry, Pollicott and others, using the method of symbolic dynamics. Here, we approach Bowen's equidistribution theory of closed geodesics through harmonic analysis on $\Gamma\backslash G$ (i.e. representation theory and trace formulae). Our method yields asymptotic expansions and sharp remainder terms which are inaccessible to dynamical methods.

The general problem of relating harmonic analysis on $\Gamma\backslash G$ to dynamical properties of the geodesic flow is of course well known. Suffice it to recall that the ergodic and mixing properties of the flow have long been studied this way. Unlike these properties, the equidistribution of closed geodesics is not a spectral invariant of the flow. Hence, it requires something more than pure representation theory, It turns out that the additional concern is the dual asymptotic behavior of the eigenfunctions of the Laplace operator as the eigenvalue tends to infinity. Due to the presence of continuous spectrum in non-compact quotients, the asymptotic behavior of eigenfunctions is a good deal more involved than in the compact case of [Z.2]. On the other hand, its study leads to results of some interest in their own right. For example, we obtain a weak (signed and averaged) version of the Lindelöf hypothesis for Rankin-Selberg zeta functions.

To state our results more precisely, let us recall some basic facts regarding the spectral decompositions of $L^2(\Gamma\backslash G)$ and $L^2(\Gamma\backslash h)$, and on the dynamics of the geodesic flow $G^t$ on the unit tangent bundle $\Gamma\backslash G$ of $\Gamma\backslash h$. Unexplained terminology or notation can be found in the appendix.

First, $L^2(\Gamma\backslash G) = {}^0L^2(\Gamma\backslash G) \oplus \Theta$, where ${}^0L^2$ is the discretely occurring subspace of cusp forms and where $\Theta$ is the subspace of incomplete $\theta$-series. In turn, $\Theta = L^2_{eis} \oplus L^2_{res} \oplus \mathbb{C}$, where $L^2_{eis}$ is the space of wave packets of Eisenstein series (of all K-weights; see below), where $L^2_{res}$ is the discretely occurring subspace of residues of Eisenstein series, and where $\mathbb{C}$ is of course the trivial representation.

Next, let $W \sim \begin{bmatrix} 0 & 1 \\ -1 & 0 \end{bmatrix}$ denote the generator of $K = SO(2)$, and let $\Omega$ denote the Casimir operator of G. An element $\sigma \in L^2(\Gamma\backslash G)$ is said to have weight m if $\frac{1}{i} W\sigma = m\sigma$. It is a Casimir eigenform of parameter s if $\Omega\sigma = s(1-s)\sigma$. We will reserve the terminology "automorphic form of weight m" for $\sigma \in C^\infty(\Gamma\backslash G)$ which are joint $(W,\Omega)$ eigenforms.

Automorphic forms of weight 0 are thus eigenfunctions of the Laplacian $\Delta$ on $L^2(\Gamma\backslash h)$. The previous decompositionof $L^2(\Gamma\backslash G)$ obviously determines a corresponding decomposition $L^2(\Gamma\backslash h) = {}^0L^2 \oplus L^2_{eis} \oplus L^2_{res} \oplus \mathbb{C}$ in weight 0. Each term has a further spectral decomposition into eigenspaces of $\Delta$. Cuspidal eigenfunctions of weight 0 will be denoted by $u_j$, so that $\Delta u_j = \lambda_j u_j$. Thus, ${}^0L^2 = \oplus\, \mathbb{C}u_j$. Aside from residual eigenfunctions (which we will temporarily ignore for simplicity), the remaining eigenfunctions are Eisenstein series $E(\cdot,\frac{1}{2}+ir)$ of $\Delta$-eigenvalue $\frac{1}{4}+r^2$. Here, E is really a vector, with one component for each cusp. For simplicity, let us also temporarily assume there is just one cusp at infinity. Then $L^2_{eis} = \int^{\oplus} \mathbb{C}\ E(\cdot,\frac{1}{2}+ir)dr$.

The spectrum of $\Delta$ thus consists of the continuum $(\frac{1}{4}, \infty)$ together with a discrete set of eigenvalues $\{0 = \lambda_0 < \lambda_1 \leq \lambda_2 \leq \cdots \uparrow\infty\}$. Following a standard notation, we will write: $\lambda_j = s_j(1-s_j) = \frac{1}{4} + r_j^2$, with $s_j = \frac{1}{2} + ir_j$. The eigenvalue $\lambda_0 = 0$ obviously corresponds to the trivial representation in $L^2(\Gamma\backslash G)$. Eigenvalues $\lambda_j \in (0, \frac{1}{4})$ correspond to complementary series irreducibles in $L^2(\Gamma\backslash G)$, and hence are called complementary series eigenvalues. For such eigenvalues, $r_j$ is pure imaginary: $r_0 = \frac{i}{2}$ and in the complementary series $r_j = it_j$ with $t_j \in (0, \frac{1}{2})$. Let $M$ be the number (possibly zero) of complementary series eigenvalues: so $t_0 = \frac{1}{2} > t_1 \geq t_2 \geq \cdots \geq t_M > \frac{1}{2}$.

Now consider the geodesic flow $G^t$ on $\Gamma\backslash G$. As is well-known, $G^t$ is given by right translation by $a_t = \begin{bmatrix} e^{t/2} & 0 \\ 0 & e^{-t/2} \end{bmatrix}$. By a closed geodesic $\bar{\gamma}$ of $\Gamma\backslash h$ one means both a closed orbit of $G^t$ and its projection to $\Gamma\backslash h$ (it should be clear from context which is meant). Each closed geodesic $\bar{\gamma}$ corresponds to a conjugacy class $\hat{\gamma}$ of hyperbolic elements of $\Gamma$ (diagonalizable over $\mathbb{R}$). To simplify notation, we will usually also confuse $\bar{\gamma}$, $\hat{\gamma}$ and elements $\gamma$ in $\hat{\gamma}$, leaving it to the context to make the meaning clear. We will write $L_\gamma$ for the length of the closed geodesic $\gamma$. Equivalently, elements of $\hat{\gamma}$ are conjugate to $\begin{bmatrix} e^{L/2} & 0 \\ 0 & e^{-L/2} \end{bmatrix}$ $(L = L_\gamma)$.

Each closed geodesic $\gamma$ determines a period orbit measure $\mu_\gamma$ on $C_b(\Gamma\backslash G)$: $\mu_\gamma(f) = \int_\gamma f$. Here, $C_b$ denotes the bounded continuous functions, and $\int_\gamma f$ is short for $\int_0^{L_\gamma} f(\gamma(t))\, dt$, $\gamma(t)$ being the natural parametrization of the orbit $\gamma$. Equidistribution theory is concerned with the weak limits of the $\mu_\gamma$ as $L_\gamma \to \infty$. To study them, it is very convenient to form the sums $\Psi_\Gamma(f,T) =$

$\sum_{L_\gamma \leq T} \mu_\gamma(f)$ and the quotients $\mu_T(f) = (\Psi_\Gamma(1,T)^{-1}\Psi_\Gamma(f,T)$. The basic result of Bowen is:

(0.1) <u>Theorem</u> ([B]) $$\lim_{T\to\infty} \mu_T(f) = \mu(f),$$

where $\mu$ is normalized Haar measure: $\mu(f) = \frac{1}{\mathrm{vol}(\Gamma\backslash G)}\int_{\Gamma\backslash G} f\, d\mu$. It is not hard to show that Bowen's theorem implies that the individual terms $\frac{\mu_T(f)}{\mu_\gamma(1)} \longrightarrow \mu(f)$ for all but a sparse subsequence (relative to counting measure on the set of lengths).

Bowen's equidistribution theorem was actually proved for Anosov flows arising one-parameter subgroups of Lie group G on a compact quotient $\Gamma\backslash G/K$ (an algebraic Anosov flow). It can be generalized to Axiom A flows, restricted to basic sets, in the form: Closed orbits are uniformly distributed relative to the measure of maximal entropy. The article [P] contains a proof, together with fuller historical remarks and references.

Our approach (as in [Z.2]) is to study the sums $\mu_T(f)$ by means of Selberg-type trace formulae. To do this, we need f to be an automorphic form in the sense above. For such forms f, we give asymptotic expansions for $\mu_T(f)$ which are reminiscent of the expansions of Huber, Selberg and many others for the case $f \equiv 1$. This is the so-called

(0.2) <u>Prime Geodesic Theorem</u>: Let $\Gamma \subset PSL_2(\mathbb{R})$ be a cofinite subgroup and let $\Psi_\Gamma(T) = \sum_{L_\gamma \leq T} L_\gamma$. Then:

$$\Psi_\Gamma(T) = \sum_{k=0}^{M} \frac{e^{s_k T}}{s_k} + O(Te^{3/4T}).$$

This theorem is analysed in detail in [He 1-2], [I] and [V]. It can surely be generalized to finite volume qoutients of rank one symmetric spaces (for some results of this kind, see [DeG], [G-W] and [S]). (No generalization to containing compact regular flats in higher rank compact quotients is known at present [B-K]).

Our generalization of (0.2) is to consider the above sums

(0.3) <u>Definition</u> $$\Psi_\Gamma(\sigma,T) = \sum_{L_\gamma \leq T} \int_\gamma \sigma$$

where $\sigma$ is an automorphic form. Our main result is:

(0.4) <u>Theorem</u> (Theorems 5A,B) Let $\Gamma \subset PSL_2(\mathbb{R})$ be cofinite. Then $\Psi_\Gamma(\sigma,T) = \sum_{k=0}^{M} \gamma_{k,m,s} < Op(\sigma)u_k,u_k> e^{s_k T} + O(e^{\frac{19}{20}T})$, for certain constants $\gamma_{k,m,s}$ and $< Op(\sigma)u_k,u_k>$.

The constants $\gamma_{k,m,s}$ are explicitly computable in terms of the weight m and Casimir-eigenvalue parameter s of $\sigma$. The constants $< Op(\sigma)u_k,u_k>$ are matrix coefficeints of an operator $Op(\sigma)$ on $L^2(\Gamma\backslash h)$ associated to $\sigma$. More precisely, $Op(\sigma)$ is the pseudo-differential operator associated to $\sigma$ in the calculus defined in [Z.6]. It is most easily introduced through its action on functions on h itself. Therefore, consider the Laplace eigenfunctions $\exp(i\lambda <z,b>)$ on h, where b is on the boundary of h, $\lambda \in \mathbb{C}$, and $<z,b>$ is the distance to i of the horocircle determined by (z,b) [He]. When $b = \infty$,

$\langle x+iy,b\rangle = \ln y$ and $\exp(i\lambda \langle z,b\rangle) = y^{i\lambda}$; the other eigenfunctions are translates of $y^{i\lambda}$ by isometries. We then define $Op(\sigma)$ as the unique operator satisfying: $Op(\sigma) \exp i\lambda \langle z,b\rangle = \sigma(z,b) \exp(i\lambda \langle z,b\rangle)$. Uniqueness follows from the Fourier inversion formula on h [He]. Also, we are regarding a function on G as a function on $h \times B$, using the natural identification between these spaces via the action of G ([He], [Z.6]).

A simple but important property of $Op(\sigma)$ is that it commutes with translation $T_g$ by an isometry if and only if $\sigma(g\cdot z,g\cdot b) = \sigma(z,b)$. Hence a $\sigma \in C_b(\Gamma\backslash G)$ defines an operator on $\Gamma\backslash h$. Unbounded functions (or, symbols) $\sigma$ can also define operators, although it requires some technicalities to prove it [Z.5].

When $\sigma$ is an automorphic form of weight m, $Op(\sigma)$ can be described in a very simple way: on an eigenfunction $u_k$ (e.g.), $Op(\sigma)u_k = \sigma L_m^- u_k$, where $L_m^-$ is the (normalized) lowering operator of $sl_2(\mathbb{R})$-theory. Thus, $L_m^- u_k$ is a form of weight (-m) and, after multiplication by $\sigma$, it goes back to weight 0 [Z.3, §1e]. All the matrix coefficients appearing in this paper involve such automorphic forms; the reader may prefer to ignore the pseudo-differential aspect of $Op(\sigma)$.

When $\sigma$ is orthogonal to 1, the matrix coefficient $\langle Op(\sigma)1,1\rangle = 0$. In that case, $\Phi_\Gamma(\sigma,T)$ is of exponentially lower order than $\Phi_\Gamma(T)$:

(0.5) <u>Theorem</u> ([6.1]) If $\sigma \perp 1$ and if the eigenvalue parameter s of $\sigma$ satisfies $\frac{1}{2} \leq \mathrm{Re} < 1$, then: $\Phi_\Gamma(\sigma,T) = O(e^{(t_1-1/2)T}) + O(e^{-1/20T})$.

This theorem leads to a proof of the equidistribution theorem $\mu_T \longrightarrow \mu$ on $C_b(\Gamma\backslash h)$ (6.4-6.5). It also gives exponential error terms for a certain class of functions $\sigma$. The problem of obtaining such error terms is also considered in [Po]. Our error terms are optimal, at least if $t_1 - 1/2 \geq 1/20$, but are only proved for somewhat special $\sigma$. We would expect such error terms to hold for all $\sigma \in C_0^\infty(\Gamma\backslash G)$ (for example), but it would be difficult to extend our results that far (cf. [Z.2, §0].

Our proof of (0.4) is based on a generalization of the Selberg trace formula (cf. [Z.2-3]). As reviewed in §1, the standard trace formula evaluates traces tr $R_\phi^\Gamma$ of convolution operators on $L^2(\Gamma\backslash G)$ in terms of the Harish-Chandra (or, conjugacy class) transform $H\varphi$. The generalization is to consider the composition $\sigma R_\phi^\Gamma$, where $\sigma$ denotes multiplication by this automorphic form. If the K-weights of $\sigma$ and $\varphi$ are adjusted properly, $\sigma R_\phi^\Gamma$ will operate on $L^2(\Gamma\backslash h)$. Tr $\sigma R_\phi^\Gamma$ can then be evaluated in terms of a generalized Harish-Chandra transform of $\varphi$, the (m,s)-parameters of $\sigma$ and the periods $\int_\gamma \sigma$. We then imitate the proofs of (0.2) which avoid the use of the Selberg zeta function. Sums of periods $\int_\gamma \sigma$ naturally replaces sums of lengths, leading to (0.4).

Some new issues inevitably arise. The most significant involve the analogues of the Weyl law which plays a key role in the proof of (0.2). This involves the eigenvalue counting function

(0.6i) <u>Definition</u> $N_\Gamma(T) = \#\{j: |r_j| \leq T\}$ and its continuous spectral analogue

$$M_\Gamma(T) = -\frac{1}{4\pi}\int_{-T}^{T} \frac{\Delta'}{\Delta}(\tfrac{1}{2}+ir)\, dr. \tag{0.6ii}$$

Here, $\Delta(s)$ is the determinant of the scattering matrix for $\Gamma$ (A10); [V, §3.5]). The Weyl law states:

(0.7) Theorem (e.g. [V, Theorem 5.2.1])

$$M_\Gamma(T) + N_\Gamma(T) = \frac{1}{2\pi} |F| \; T^2 + O(T \ln T),$$

where $|F|$ is the area of a fundamental domain $F$ for $\Gamma$.

The proof of the analogues of (0.7) for $\sigma \neq 1$ first involves the sum function:

(0.8) Definition $N_\Gamma(\sigma,T) = \sum_{|r_j|\leq T} \langle Op(\sigma)\, u_j,\, u_j \rangle.$

Here, the matrix coefficients $\langle Op(\sigma)\, u_j,\, u_j\rangle$ are well defined for any automorphic form $\sigma$ (cusp form, Eisenstein series or residue) due to the rapid decay of $u_j$ in the cusps. The continuous analogue $M_\Gamma(\sigma,T)$ is harder to define. When $\sigma$ is a cusp from, the matrix coefficients $\langle Op(\sigma)\, E(\cdot,\frac{1}{2}+ir),$ $E(\cdot,\frac{1}{2}+ir)\rangle$ are well-defined and continuous in r, and we may set

(0.9) Definition $M_\Gamma(\sigma,T) = - \frac{1}{4\pi} \int_{-T}^{T} \langle Op(\sigma)\, E(\cdot,\frac{1}{2}+ir),\, E(\cdot,\frac{1}{2}+ir)\rangle \, dr$

(for cuspidal $\sigma$).

When $\sigma$ is an Eisenstein series, the matrix coefficient is not well-defined. We will follow Zagier's method of renormalizing inner products [Za 2] to give it a meaning . We therefore replace $\langle Op(\sigma)\, E(\cdot,\frac{1}{2}+ir),\, E(\cdot,\frac{1}{2}+ir)\rangle$ by its renormalization $RN\langle Op(\sigma)\, E(\cdot,\frac{1}{2}+ir),\, E(\cdot,\frac{1}{2}+ir)\rangle$, whose precise definition and properties will be discussed in §4. For the moment, we only note that, if $\sigma = E(\cdot,s)$, it coincides with a Rankin-Selberg convolution $R(|E(\cdot,\frac{1}{2}+ir)|^2,s)$ (cf.

(A13(c))); while if $\sigma \equiv 1$, it coincides with $\frac{\Delta'}{\Delta}(\frac{1}{2}+ir)$. Thus,

(0.10) <u>Definition</u> $M_\Gamma(\sigma,T) = -\frac{1}{4\pi}\int_{-T}^{T} RN\langle Op(\sigma)\; E(\cdot,\frac{1}{2}+ir),\; E(\cdot,\frac{1}{2}+ir)\rangle\, dr.$

The analogues of the Weyl law for cusp forms is:

(0.11) <u>Theorem</u> (4A) $(M_\Gamma + N_\Gamma)(\sigma,T) = O(T/\ln T)$ ($\sigma$ cuspidal).

The proof of (0.11) is very similar to that in the case of co-compact $\Gamma$ ([Z.2, §4]).

The Weyl law for Eisenstein series $\sigma = E(\cdot,s)$ is quite different. Using [D-I], we show:

(0.12) <u>Theorem</u> (4B) $(M_\Gamma + N_\Gamma)(E(\cdot,s),\, T) = O(T^{3/2})$ $\quad \mathrm{Re}\, s = 1/2.$

The estimate of (0.12) is of some interest in its own right, since it gives a partial confirmation of the mean Lindelöf hypothesis (MLH) for Rankin-Selberg zeta functions ([I.1, p.139; I.2, p.188).

(0.13) Conjecture (MLH) For $\Gamma = SL_2(\mathbb{Z})$ and $\mathrm{Re}\, s = 1/2$,

$$\sum_{|r_j|\le T} |\langle E_s u_j,\, u_j\rangle| \ll |s|^A\, T^{3/2+\epsilon}$$

For $\Gamma = SL_2(\mathbb{Z})$, the $M_\Gamma$ term is known to be of lower order than $N_\Gamma$, so that (0.13) is just the conjecture that the estimate in (0.12) holds for the associated absolute sum. The transition from (0.12) to (0.13) seems quite difficult. In [Z.5] we show at least that the sum in (0.13) is $o(T^2)$. Our method could give also the slightly better result $O(T^2/\ln T)$.

The Weyl laws (0.11) and (0.12) are crucial in the proof of (0.4). Since (0.12) is of higher order than (0.11), it should not be surprising that the remaninder in (0.4) is of higher order than in (0.2). This large remaninder is also due our crude estimate in the growth of the periods $\int_\gamma E(\cdot,s)$ as $L_\gamma \to \infty$ ($\mathrm{Re}\, s = 1/2$). Based on the fact that a closed geodesic can only go a distance of $e^{\frac{1}{2}L_\gamma}$ away from a compact part of $\Gamma\backslash h$ into an end, we bound $\int_\gamma E(\cdot,s)$ by $O(e^{\frac{1}{4}L_\gamma})$ It is quite likely that a typical closed geodesic only goes a distance $L_\gamma$ into an end (cf. [Su]); so this aspect of the remainder estimate could probably be improved a great deal.

As mentioned above, this paper is an extension of our earlier work on a compact hyperbolic surface [Z.2]. The results of that work will be summarized in §1, but will otherwise be assumed without proof. Thus, this paper is far from self-contained.

Although we only consider hyperbolic surfaces in [Z.2] and in this paper, it seems reasonable that similar asymptotic expansions should exist on a general cofinite quotient of a rank one symmetric space. In fact, it seems likely that they should occur on any locally symmetric quotient for which a prime geodesic theorem could be proved. In particular, it seems reasonable to predict that if a prime geodesic theorem could be proved for compact regular flats in higher rank quotients, then these flats could be shown by our more general trace formulae to be equidistributed. However, the analysis would have to be very long and clumsy, since it would involve a case-by-case study of automorphic forms.

Acknowledgements I. Efrat and P. Sarnak pointed out to us several years ago that Zagier [Za.1] had already developed a well-known trace formula for Tr $E(\cdot,s)R_{\varphi}^{\Gamma}$, at least if $\Gamma = SL_2(\mathbb{Z})$. Zagier's method was clearer and more efficient than our former "truncation" method, and we have therefore adopted it in §3. We also use Zagier's renormalization method [Za.2] at various points here and in [Z.5]. Sarnak's proof of the prime geodesic theorem ([Sa.1]), which avoids the use of the Selberg zeta function, was also indispensible.

## § 1. Review of [Z.2] on the compact case

At the end of this paper, the reader will find an appendix containing a long list of notations and definition. Most are relatively standard and follow [F], [Hej], [L], [Za1-2], [V]. A few, in particular those involving $\psi$DO's or Harish-Chandra type transforms, are non-standard and follow our earlier articles [Z2-3]. Reference numbers of the form (A10) (e.g) refer to this appendix.

Our purpose here is to give a brief resume of the results of [Z2]-[Z3] on generalized trace formulae, Weyl laws and prime geodesic theorems for the compact case. These results will play a crucial role in § 5. We also indicate the kinds of modifications (many quite substantial) which are required to extend them to the present cofinite case.

The basic technique of this paper, as in [Z2]-[Z3], is a detailed analysis of certain trace formulae. These trace formulae generalize the classical formula of Selberg for the trace of a convolution operator $R_\phi^\Gamma$ (A14). The generalization is just to consider the operators $\sigma R_\phi^\Gamma$, i.e. $R_\phi^\Gamma$ followed by multiplication by an automorphic form $\sigma$. When $\phi \in S_{m,0}$ (A14) and when $\sigma$ has weight m, $\sigma R_\phi^\Gamma$ acts on the space of forms of weight 0 or, equivalently, on $L^2(\Gamma\backslash h)$. The operator $\sigma R_\phi^\Gamma$ can then be identified as a composition $Op(\sigma)h(R)$, where $\Delta = -(\frac{1}{4}+R^2)$, where $h(R)$ is introduced in (A14), and where $Op(\sigma)$ (A12) is the pseudo-differential operator ($\psi$DO) with complete symbol $\sigma$ in the sense of [Z1]-[Z3]. We refer to [Z3], § 1(d), for background on the $\psi$DO's $Op(\sigma)$, and to [Z1], § 1(B) for further discussion of the passage from operators of the form $\sigma R_\phi^\Gamma$ to those of the form $Op(\sigma)h(R)$. In general, the matrix elements $(Op(\sigma)u_k, u_k)$ are best studied through the traces tr $Op(\sigma)h(R)$, while the geodesic periods $\int_\gamma \sigma$ are best studied through the traces tr $\sigma R_\phi^\Gamma$.

In the compact case, the trace formulae have the form:

$$\text{Tr } \sigma R_\phi^\Gamma = (\text{spec}) = (\text{hyp}), \tag{1.1}$$

where

$$(\text{spec}) = \sum_j (\text{Op}(\sigma)u_j, u_j) S_m \phi(s_j) \tag{1.2}$$

(see [Z2], § 2) and where

$$(\text{hyp}) = \sum_{\{\gamma\}_{\text{hyp}}} (\textstyle\int_{\gamma_0} \sigma) \, H_{s,m} \phi(a(\gamma)). \tag{1.3}$$

Here $H_{s,m}$ is (as in A16(b)) a non-standard HC transform and (s,m) are the $(\Omega, W)$ parameters of $\sigma$ (A1). More precisely, (1.3) is correct for weight 0 or discrete series $\sigma$ but equals just a kind of imaginary part for higher weight, continuous series forms (cf. [Z2], Proposition 2.11; [Z3], Theorem 5.2). Note that the identity class term vanishes, since $\sigma$ is orthogonal to the constant functions.

Similar trace formulae hold for operators of the form $\text{Op}(\sigma)h(R)$. Actually, it is very useful to consider the slightly more general class of operators $\text{Op}(\chi\sigma)h(R)$, where $\chi$ is a cut-off function on $h \times h$, supported in a neighborhood of the diagonal (see [Z3], § 1(f)). (1.2) and (1.3) then become:

$$(\text{spec}) = \sum_j (\text{Op}(\sigma)u_j, u_j) J_m \chi(r_j) h(r_j) \tag{1.2'}$$

(1.3') $$(\mathrm{hyp}) = \sum_{\{\gamma\}_{\mathrm{hyp}}} (\textstyle\int_{\gamma_0} \sigma)\, M^{\chi}_{s,m} \hat{h}(L_\gamma).$$

Here, $J_m\chi$ is a certain transform of $\chi$, with $J_m\chi(r) = 1 + O(r^{-N})$ $(\forall N)$, as $|r| \to \infty$ ([Z3], § 1(f)); and $M^{\chi}_{s,m}$ are certain operators depending only on the displayed parameters ([Z3], § 5).

The evaluations (1.2)-(1.2') and (1.3)-(1.3') are given in [Z2]-[Z3] and will be assumed without proof in this paper. Perhaps, though, it would be helpful to sketch the main point. First, the kernel $L(x,y)$ of $\sigma R^{\Gamma}_{\phi}$ has the form: $L(x,y) = \sigma(x)K(x,y)$, where

(1.4) $$K(x,y) = \sum_{\gamma\in\Gamma} \phi(x^{-1}\gamma y).$$

For $\phi \in C^{\infty}_0(G)$, (1.4) is a finite sum and following a standard argument, the terms may be re-arranged in conjugacy classes. This leads to the kernels

(1.5) $$K_{\{\gamma\}}(x,y) = \sum_{\sigma\in\Gamma_\gamma\backslash\Gamma} \phi(x^{-1}\sigma^{-1}\gamma\sigma y).$$

Clearly $K_{\{\gamma\}}(x,x)$ is $\Gamma$-invariant and of weight $-m$ if $\sigma \in S_{m,0}$. So $L_{\{\gamma\}}(x,x) = \sigma(x)K_{\{\gamma\}}(x,x)$ has weight 0, and its integral over $\Gamma\backslash G$ is the inner product $\langle\sigma,K_{\{\gamma\}}\rangle$. The key identity leading to (1.3) is:

(1.6) $$\langle\sigma,K_{\{\gamma\}}\rangle = (\textstyle\int_{\gamma_0}\sigma)H_{m,s}\phi(a(\gamma)) + (\textstyle\int_{\gamma_0}(X_+\sigma))G_{m,s}\phi(a(\gamma)).$$

Here $G_{m,s}$ is another HC transform, like $H_{m,s}$ but involving another special function. The second term vanishes if $\sigma$ has weight 0, or if it is from the discrete series; and can be made to vanish by taking an appropriate kind of imaginary part, to be denoted by *Im* (see [Z2], § 2(II); the notation "*Im*" is not used there, however). Formula (1.3) then follows from (1.6). Finally, (1.2) is a straightforward consequence of the spectral decomposition.

Two main applications of the trace formulae are generalized Weyl laws and prime geodesic theorems. The Weyl laws for the compact case state ([Z3], § 5):

$$(1.7) \qquad N_\Gamma(\sigma,T) \overset{\text{def}}{=} \sum_{|r_j|\le T} (Op(\sigma)u_j,\ u_j) \ll_{(s,m)} T/\ln T$$

(where (s,m) are as usual the (Ω,W) parameters of $\sigma$).

The prime geodesic theorems for the compact case state ([Z2], § 4):

$$(1.8) \quad \psi_\Gamma(\sigma,T) \overset{\text{def}}{=} \sum_{L(\gamma)\le T} \int_{\gamma_0} \sigma = \sum_{k=1}^{M} \langle Op(\sigma)u_k,\ u_k\rangle \gamma_{k,s,m}\, e^{(\frac{1}{2}+t_k)T} + O_{s,m}(Te^{3/4T}).$$

Here $\gamma_{k,s,m}$ are certain explicit constants ([Z2], § 3) which will be described further in § 6.

The Weyl laws (1.7) are required in the proof of the prime geodesic theorems (1.8). Also required is an estimate in the absolute sums

$$(1.9) \qquad |N_\Gamma(\sigma,T)| \overset{\text{def}}{=} \sum_{|r_j|\le T} |\langle Op(\sigma)u_j,\ u_j\rangle|.$$

The obvious estimate

$$|N_\Gamma(\sigma,T)| \underset{(s,m)}{<<} N_\Gamma(T) \tag{1.10}$$

is all that one needs. The generalization of (1.7) and (1.8) to the finite area case will be the object of § 4; perhaps surprisingly, even the extension of (1.10) is non-trivial.

Generalization of the prime geodesic theorems (1.8) to the finite area case is the primary goal of this paper. Much of what occurs in the compact case carries over verbatim to the general finite area case. This applies in particular to the hyperbolic and discrete spectral terms of the trace formula. Note that (1.8) is, up to the error term, a relation between the hyperbolic terms and the complementary series spectral terms. Since the latter always occur discretely, one expects (1.8) to be valid (up to the error term) in the finite area case. This indeed proves to be the case.

The analysis of the hyperbolic and discrete spectral terms in [Z2] will be relied on heavily in this paper. For the sake of completeness and intelligibility, let us summarize here the main steps leading to (1.8).

The first step is to invert the non-standard HC transforms arising in the hyperbolic terms of the trace formula (1.1). Setting $\psi = H_{m,s}\phi$, where $\phi \in C_0^\infty$, the trace formula in the compact case becomes:

$$\sum_j (Op(\sigma)u_j,u_j) S_m \circ H_{m,s}^{-1}\psi(s_j) = \sum_{\{\gamma\}_{hyp}} (\textstyle\int_{\gamma_0}\sigma)\psi(L_\gamma). \tag{1.11}$$

For concreteness, let us assume $m = 0$ (the other cases are similar). The transform $S_m \circ H_{m,s}^{-1}$ can then be given the quite explicit expression:

$$S_m \circ H_{m,s}^{-1} \overset{\text{def}}{=} M_s^c, \tag{1.12a}$$

where, for $\psi \in C_0^\infty(\mathbb{R}^+)$,

$$(1.12b) \qquad \mathbf{M}_s^c\psi(s_0) = \int_{\mathrm{Re}\ s=\tau} \mathbf{M}\psi(s)[\Gamma_k^+(s,s_0) + \Gamma_k^-(s,s_0)]\frac{ds}{i}.$$

Here $\tau > \frac{1}{2}\mathrm{Re}(s_0+1)$, $\mathbf{M}$ is the usual Mellin transform on $\mathbb{R}^+$, and

$$(1.13) \qquad \Gamma_k^\pm(s,s_0) = \frac{2^{s-(3/2)}\Gamma(s)\Gamma(s-\frac{1}{2})\Gamma(\pm\frac{1}{2}s_0+s-\frac{1}{2})\Gamma(-2s+2)}{\Gamma(s+\frac{1}{4}s_k-\frac{1}{4})\Gamma(s-\frac{1}{4}s_k-\frac{1}{4})\Gamma(\pm\frac{1}{2}s_0-s+\frac{3}{2})}$$

(see [Z2], Proposition 3.3).

Suppose temporarily that (1.11) is valid for $\psi \in C_0^\infty(\mathbb{R}^+)$; we will return to this question at the end of this section. Then let $\psi_T$ be the characteristic function of the length interval $[1,T]$ and $\psi_{T,\epsilon} \in C_0^\infty$ an appropriate smoothing. We substitute $\psi_{T,\epsilon}$ into the trace formula (1.11). As usual, the complementary sreies are exponentially growing as $T \longrightarrow \infty$, while the principal series terms are oscillatory. The latter may be estimated by the remainder term in (1.8). The former contribute a finite sum of integrals involving $\mathbf{M}_s^c\psi_{T,\epsilon}$. These may be evaluted very explicitly, thanks to (1.12b). If we deform the line of integration leftwards, we pick up some residual terms

$$(1.14) \qquad \mathrm{Res}_{s=\frac{1}{2}+s_j} (\Sigma_\pm\Gamma_k^\pm(s,s_j))\mathbf{M}\psi_{T,\epsilon}(s).$$

Combining (1.14) with the simple asymptotic

$$(1.15) \qquad \mathbf{M}\psi_{T,\epsilon}(\tfrac{1}{2}+s_j) \sim (\tfrac{1}{2}+s_j)^{-1}e^{T(\frac{1}{2}+s_j)},$$

we get (1.8). For further details on the derivation of (1.8), particularly on the estimation of error terms, we refer to [Z2].

The main problem in generalizing (1.8) to finite area is to deal with the new features in the trace formulae. The operators $\sigma R_\phi^\Gamma$ or $Op(\chi\sigma)h(R)$ are now traceable only on the space ${}^0L_\chi^2$ of cusp forms (A9, A17). The trace formula then takes the preliminary shape:

$$\text{(1.16)} \qquad \operatorname{Tr} \pi_0 \sigma R_\phi^\Gamma \pi_0 = (\text{hyp}) + (\text{ell}) + (\text{III}) ,$$

where (ell) is a finite sum over elliptic conjugacy classes and where (III) is a combination of parabolic and continuous spectral terms. It is most illuminating to view (III) as a spectral term, contributing an integral of the form

$$\text{(1.17)} \qquad (\text{III}) = \int_{\infty}^{\infty} \text{RN}\langle Op(\sigma)E(\cdot,\tfrac{1}{2}+ir),\ E(\cdot,\tfrac{1}{2}+ir)\rangle S\phi(\tfrac{1}{2}+ir)dr$$

to the discrete spectral term on the left of (1.16). Here $\langle Op(\sigma)E(\cdot,\frac{1}{2}+ir),\ E(\cdot,\frac{1}{2}+ir)\rangle$ does not actually make sense if $\sigma$ is an Eisenstein series, a constant or an eigenfunction of the residual spectrum. The notation "RN" in (1.17) stands for "renormalization," or, in other words, for the finite part of the divergent integral (the terminology is due to Zagier; the rigorous version will be given in § 2). From this point of view, (1.17) really has the familiar form (spec) = (hyp), up to some very minor terms that can be analyzed as in the classical case where $\sigma \equiv 1$.

The subtlest terms arise from the finite parts of the inner products $\langle Op(\sigma)E(\cdot,\frac{1}{2}+ir),\ E(\cdot,\frac{1}{2}+ir)\rangle$. These give rise to the functions $M_\Gamma(\sigma,\chi,T)$

mentioned in the introduction:

$$\text{(1.18a)} \qquad M_\Gamma(\sigma,\chi,T) = -\frac{1}{4\pi}\int_{-T}^{T} \langle Op(\sigma)E(\cdot,\tfrac{1}{2}+ir),\ E(\cdot,\tfrac{1}{2}+ir)\rangle dr$$

(if $\sigma$ is cuspidal)

$$\text{(1.18b)} \qquad = -\frac{1}{4\pi}\int_{-T}^{T} R[|E(\cdot,\tfrac{1}{2}+ir)|^2,\ \tfrac{1}{2}+ir]dr$$

(if $\sigma = E(\cdot,\frac{1}{2}+ir)$, e.g.)

Much of the analysis of this paper is devoted to estimating the contribution of these continuous spectral objects to Weyl laws and to the error of the prime geodesic theorem.

Finally, we would like to discuss a technical point raised above. Namely, some of our applications of the trace formula require the use of non-compactly supported test functions. Most importantly, these test functions occur when we use the version (1.11) of the formula. Here $\psi \in C_0^\infty(\mathbb{R}^+)$, but $\phi = H_{s,m}^{-1}\psi \notin C_0^\infty$. Of couse, it is quite commomplace to use test functions $\phi \notin C_0^\infty$, but the validity of (1.11) must be established for them. The main point, as will be discussed in detail in § 5, is to justify the trace formulae for $\psi \in C_0^\infty(\mathbb{R}^+)$ by a continuity argument. Namely, we view the trace formula as an identity between linear functionals on $C_0^\infty$. We show that these functionals are continuous on a certain Banach space in which $C_0^\infty$ is dense. The test functions $\phi$ which are required in applications will turn out to lie in this space. This argument, implicit in [Z2], § 4, will be made explicit in § 5.

## § 2. Trace formula for Tr $\sigma R_\varphi^\Gamma$: $\delta \in {}^0L^2(\Gamma\backslash G)$

In this section we give explicit formulae for Tr $\pi_0 \sigma R_\varphi^\Gamma \pi_0$ when $\sigma$ is cuspidal. This is straightforward since the rapid decay of $\sigma$ in the cusps implies that $\sigma R_\varphi^\Gamma$ is actually traceable on all of $L^2(\Gamma\backslash G)$.

The resulting formula depends on the $(\Omega,W)$ parameters of $\sigma$, so the section is organized accordingly. Throughout, we will be looking at $\mathrm{Tr}_\chi \sigma R_\varphi^\Gamma$, the trace of $\sigma R_\varphi^\Gamma$ on $L_\chi^2(\Gamma\backslash G)$. Recall that $\{u_j\}$, $\{E_a(\cdot,s)\}$, $\{\theta_{aj}^\chi \overset{\text{def}}{=} \theta_a(\cdot,s_j)\}$, $\{\psi_n\}$ denote fixed ONB's for ${}^0L^2 \cap H_0$, $L_{eis}^2 \cap H_0$, $L_{res}^2 \cap H_0$ and the lowest weight vectors of discrete series irreducibles, respectively. To save space we will abbreviate the eigenvalues $s_j(\chi)$ and ONB's $u_j(z,\chi)$ by $s_j^\chi$ and $u_j^\chi$ (similarly for the others).

(A) $\sigma \in {}^0L^2(\Gamma\backslash G) \cap H_0$, $\varphi \in C_0^\infty(G//K)$. Let $(s_k,0)$ be the $(\Omega,W)$ parameters of $\sigma$; so $\sigma = u_k$.

Theorem 2(A)

$$\mathrm{Tr}_\chi u_k R_\varphi^\Gamma = \sum^{h(\chi)} \sum_{j=1}^{M(a)} \langle u_k \theta_{aj}^\chi, \theta_{aj}^\chi \rangle S\varphi(s_j^\chi) + \sum_{n=1}^{\infty} \langle u_k u_n^\chi, u_n^\chi \rangle S\varphi(s_n^\chi)$$

$$+ \sum_{a=1}^{h(\chi)} \frac{1}{4\pi} \int_{\infty}^{\infty} \langle u_k E_a(\cdot,\tfrac{1}{2}+ir,\chi), E_a(\cdot,\tfrac{1}{2}+ir,\chi) \rangle S\varphi(s_n^\chi)$$

$$= (\mathrm{hyp}) + (\mathrm{ell}),$$

where S is the spherical transform,

$$(\text{hyp}) = \sum_{\{\gamma\}_{\text{hyp}}} \chi(\gamma)(\textstyle\int_{\gamma_0} u_k) H_k^c\ \varphi(a(\gamma)$$

$$(\text{ell}) = \sum_{\{\gamma\}_{\text{ell}}} \chi(\gamma)\frac{1}{\nu(\gamma_0)} u_k(z_\gamma) L_k^c\ \varphi(k(\theta_\gamma)).$$

$(\gamma(\gamma_0)$ = order of $\gamma_0)$.

Proof The left hand side follows immediately from the spectral decomposition of $L^2_\chi(\Gamma\backslash G)$. It is clear that all sums and integrals converge absolutely.

The right hand side is $\int_{\Gamma\backslash G} K(x,x,\chi)u(x)dx$ where $K(x,y,\chi)$ is the kernel of $R^\Gamma_\varphi$ on $L^2_\chi$.

Breaking up $K(x,x,\chi) = K_{(e)} + \sum_{\{\gamma\}_{\text{hyp}}} \chi(\gamma) K_{\{\gamma\}}(x,x) + \sum_{\{\gamma\}_{\text{ell}}} \chi(\gamma) K_{\{\gamma\}}(x,x)$

$+ \sum_{\{\gamma\}_{\text{par}}} \chi(\gamma) K_{\{\gamma\}}(x,x)$ (as explained in § 1 and Appendix), we get correspondingly the terms (e) + (hyp) + (ell) + (par) for the trace. Clearly, (e) = $\varphi(e)\ \langle u_k, 1\rangle = 0$ and (par) = 0 since $K_{\{\gamma\}}(x,x) \in \mathcal{B}(\Gamma\backslash G)$ (see § 1 and Appendix).

On the other hand,

$$\int_{(\Gamma\backslash G)} K_{\{\gamma\}}(x,x) u_k(x) dx = H_k^c\ \varphi(a(\gamma)) \textstyle\int_{\gamma_0} u_k$$

exactly as in [Z2].

For elliptic $\gamma$, let $a_\gamma \in PSL_2(\mathbb{R})$ conjugate $\gamma$ to $k(\gamma)$: $a_\gamma^{-1}\gamma a_\gamma = k(\gamma)$. Then

$$I_\gamma(u_k,\varphi) \overset{\text{def}}{=} \int_{(\Gamma\backslash G)} K_{\{\gamma\}}(x,x) u_k(x) dx$$

$$= \int_{G_\gamma\backslash G} \varphi(x^{-1}\gamma x)\,[\int_{\Gamma_\gamma\backslash G_\gamma} u_k(g_1 x)\,dg_1]\,dx$$
$$= \int_{K\backslash G} \varphi(x^{-1}k(\gamma)x)\,[\int_{\langle k(\gamma_0)\rangle\backslash K} u_k(a_\gamma^{-1}kx)\,dk]\,dx$$

where $\langle k(\gamma_0)\rangle$ is the finite cyclic group generated by the primitive element $k(\gamma_0)$. Using the KAK dcomposition of G we get:

$$I_\gamma(u_k,\varphi) = \int_A \varphi(x^{-1}k(\gamma)x)\;[\int_{\langle k(\gamma_0)\rangle\backslash K} u_k(a_\gamma^{-1}ka)\,dk]\,d\mu(a)$$

where $d\mu(a)$ is as above (cf. [L], p. 139).

Let $I_\gamma(u_k,a_t) = \int_{\langle k(\gamma_0)\rangle\backslash K} u_k(a_\gamma^{-1}ka_t)\,dk \qquad [a_t = \begin{bmatrix} e^{t/2} & 0 \\ 0 & e^{-t/2}\end{bmatrix}]$.

Recall that the spherical means operator $M_t$ on $H_0$ is:

$$M_t u(x) = \int_K u(xka_t)\,\frac{dk}{2\pi}.$$

It follows that:

$$I_\gamma(u_k,a_t) = \frac{2\pi}{\nu(\gamma_0)} M_t u_k(a_\gamma^{-1}\cdot i) = \frac{2\pi}{\nu(\gamma_0)} M_t u_k(z_\gamma), \quad (z_\gamma = \text{fixed point of } \gamma_0).$$

Now on symmetric spaces, $M_t$ is a function of $\Delta$. Indeed, if $\Delta u = -(\lambda^2+1)u$ then $M_t u = \varphi_\lambda(a_t\cdot 0)$, $\varphi_\lambda$ being the spherical function [He].

Consequently: $I_\gamma(u_k,a_t) = \frac{1}{\nu(\gamma_0)}\varphi_{r_k}(a_t\cdot 0)u_k(z_\gamma)$, and

$$I_\gamma(u_k,\varphi) = \frac{1}{\nu(\gamma_0)} u_k(z_\gamma)\int_A \varphi(a^{-1}k(\gamma)a)\varphi_{r_k}(a\cdot 0)\,d\mu(a). \qquad \blacksquare$$

(B) $\sigma \in {}^0L^2 \cap H(\pm 2)$, $\varphi \in C_0^\infty \cap S_{2,0}$, $\sigma$ in the continuous series.

Thus: $\sigma = E^\pm u_k$, $E^\pm = H \pm iV$.

As in [Z2] we will consider the difference traces (the "imaginary parts"):

$$\mathrm{Tr}_\chi(E^+u_kR_\varphi - E^-u_kR_{\varphi^*}) \quad \text{where}$$

$$\varphi^*(x) = \int_{\mathrm{Re}s=0} g(s)\phi_{-2,s}(x)\,\frac{ds}{i} \quad \text{if} \quad \varphi(x) = \int_{\mathrm{Re}s=0} g(s)\phi_{2,s}(x)\,\frac{ds}{i}$$

($\phi_{m,s}$ are the generalized spherical functions).

Theorem 2(B)

$$\mathrm{Tr}_\chi(E^+u_kR^\Gamma_\varphi - E^-u_kR^\Gamma_{\varphi^*})$$
$$= \sum_{\alpha=1}^{h(\chi)} \sum_{j=i}^{M(\alpha)} \langle \mathrm{Op}(X_+u_k)\,\theta^\chi_{\alpha j},\ \theta^\chi_{\alpha j}\rangle S_2\varphi(s_j) + \sum_{n=1}^{\infty} \langle \mathrm{Op}(X_+u_k)\,u^\chi_n,\ u^\chi_n\rangle\ S_2\varphi(s^\chi_n)$$
$$+ \sum_{\alpha=1}^{h(\chi)} \frac{1}{4\pi}\int_{-\infty}^{\infty} \mathrm{Op}(X_+u_k)E_\alpha(\cdot,\ \tfrac{1}{2}+ir,\ \chi),\ E_\alpha(\cdot,\ \tfrac{1}{2}+ir)\rangle\ S_2\varphi(\tfrac{1}{2}+ir)\,dr$$
$$= (\mathrm{hyp})$$

where $(\mathrm{hyp}) = \sum_{\{\gamma\}_{\mathrm{hyp}}} \chi(\gamma)(\int_{\gamma_0} X_+u_k)H^c_{k,+}\varphi(v_\gamma),\ v_\gamma = \frac{1}{2}(\mathrm{sh}L_{\gamma/2})^2.$

Proof The spectral side and hyperbolic terms work out exactly as in the compact case ([Z2]), modulo the obvious modifications required by the presence of $L^2_{\mathrm{eis}}$.

As in case (A), the identity term (e) and parabolic term vanish. We have only to show (ell) = 0. Now (ell) $= \sum_{\{\gamma\}} \chi(\gamma)I^{\mathrm{odd}}_\gamma(E^+u_k,\varphi)$, where
$I^{\mathrm{odd}}_\gamma(E^+u_k,\varphi) = \frac{1}{2i}[I_\gamma(E^+u_k,\varphi) - I_\gamma(E^-u_k,\varphi^*)]$, and
$I_\gamma(\sigma,\varphi) = \int_{G_\gamma\backslash G} \varphi(x^{-1}\gamma x)\,[\int_{\Gamma_\gamma\backslash G_\gamma} \sigma(g_1x)dg_1]dx.$

We claim $I^{\mathrm{odd}}_\gamma(E^+u_k,\varphi) = 0$ if $\gamma$ is elliptic. Indeed, if $\varphi \in S_{2,0}$ one has:
$I_\gamma(E^+u_k,\varphi) = \int_A \varphi(a^{-1}k(\gamma)a)\ [\int_{\langle k(\gamma_0)\rangle\backslash K} (E^+u_k)(a_\gamma^{-1}ka)dk]\,d\mu(a).$

It follows easily as in [Z2] that

$$I_\gamma^{odd}(E^+u_k,\varphi) = \int_A \varphi(a^{-1}k_\gamma a)\ [\int_{<k(\gamma_0)>\backslash K}(X_+u_k)(a_\gamma^{-1}ka)dk]d\mu(a).$$

The inner integral is

$$\frac{1}{\nu(\gamma_0)}\int_K X_+u_k(a_\gamma^{-1}ka)\ dk$$

where (hyp) $= \sum_{\{\gamma\}_{hyp}} \chi(\gamma)(\int_{\gamma_0}\psi_m)H_m^d\ \varphi(v_\gamma)$ where $\gamma \sim \begin{bmatrix} a & 0 \\ 0 & a^{-1}\end{bmatrix}$, $w = \dfrac{a+a^{-1}}{a-a^{-1}}$, $V_\gamma = w^{-2}$, $H_m^d\ \varphi(v) = wv^{m/4}\int_{-\infty}^{\infty}(u+i)^{-m/2}[1+\frac{v-1}{u^2+1}]^{-m/4}\varphi^D\ [1+\frac{v-1}{u^2+1}]\ du$ ($\varphi^D$ is the induced function on the unit disc $\approx G/K$).

<u>Proof</u> The spectral side, (e), (par), and (hyp) work out as in [Z2] and as in case B. As in case B, (ell) = 0 but here the proof is easier: $I_\gamma(\psi_m,a_t) = M_t\psi_m(a_\gamma^{-1}) = 0$ since there are no K-invariant vectors in the irreducible determined by $\psi_m$. ∎

## § 3. Trace Formula for $\mathrm{Tr}\sigma R_\varphi^\Gamma$: $\sigma \in \Theta$

Our purpose in this section is to work out trace formulae for $\mathrm{Tr}\pi_0\sigma R_\varphi^\Gamma\pi_0$, where $\sigma$ is one of the generating forms (over $A$) for $\Theta$; i.e. $\sigma = \{E(z,s),\ X_+E(g,s),\ \text{residue}\}$.

As mentioned in the introduction, we will follow Zagier's approach in [Za1] and compute $\mathrm{Tr}E_sR_\varphi^\Gamma$ for $\mathrm{Re}s > 1$ by the unfolding method. We are then required to analytically continue the resulting terms. Here we cannot, unfortunately, refer to [Za1] since only the group $\Gamma = SL_2(Z)$ is considered there. The extra difficulties in dealing with general co-finite $\Gamma$ arise at least because (a) such functions as $R_{\alpha\beta}(t,s)$ (see A13c) are not arithmetic zeta functions and hence require non-classical analytic continuation arguments and (b) the existence of residual spectrum in $(1/2,1)$ complicates a contour shifting argument. None of these problems is very serious, and known methods can be used to take care of them; but they will lengthen this section somewhat.

Recall (from the Appendix) that: $\{\kappa_\alpha\}$ denotes a complete set of inequivalent cusps; $\sigma_\alpha \in G$ is the element so that $\sigma_\alpha\cdot\infty = \kappa_\alpha$ and $\sigma_\alpha^{-1}\Gamma_\alpha\sigma_\alpha = \Gamma_\infty = \{\begin{pmatrix}1 & n\\ 0 & 1\end{pmatrix},\ n \in z\}$; for a given character $\chi: \Gamma \to U(1)$, $h(\chi)$ denotes the number of the $\{\kappa_\alpha\}$ so that $\chi|_\Gamma \equiv 1$; $\pi_0$, $\pi_{res}$, $\pi_{eis}$, $\pi_1$ denote the orthogonal projections onto ${}^0L^2_\chi$, ${}^0L^2_{\chi,eis}$, ${}^0L^2_{\chi,res}$ and $\mathbb{C}$ (if $\chi \equiv 1$ on $\Gamma$); $K_0$, $K_{eis}$, $K_{res}$ and $(K1,1)$ are the kernels of $R_\varphi$ on these spaces.

(A) $\sigma = E_s$, $\varphi \in C_0^\infty(G//K)$.

Set: $I_\alpha(s,\chi,\varphi) = \mathrm{Tr}_\chi\pi_0E_\alpha(\cdot,s)R_\varphi^\Gamma\pi_0 = \int_F E_\alpha(z,s)K_0(z,z)dz$. Since $K_0(z,z) = \sum_{j=1}^\infty |u_j(z)|^2h(r_j)$ is rapidly decaying in all cusps (with our assumption on $\varphi$), $I_\alpha(s,\chi,\varphi)$ is clearly meromorphic on $\mathbb{C}$ with poles among the poles of $E_a(z,s)$. $I(z,s,\varphi) = [I_1(z,s,\varphi), \cdots, I_{h(\chi)}(z,s,\varphi)]$ also inherits a functional equation from that of E:

(a) $E(z,1-s,\chi) = \Phi(1-s,\chi)E(z,s,\chi)$

(b) $\Phi(1-s,\chi)\Phi(s,\chi) = I$

(c) $\Phi^t(s,\chi) = \Phi(s\ \bar{\chi})$.

Here $E = (E_1, \cdots, E_{h(\chi)})$, and $\Phi(s,\chi) = [\Phi_{\alpha\beta}(s,\chi)]_{\alpha,\beta=1,\cdots,h(\chi)}$. $I(s,\chi,\varphi)$ has the functional equation of $E(z,s,1)$.

The trace formula is given by:

Theorem 3(A)

$$I_a(s,\chi,\varphi) = \sum_{j=1}^{\infty} \langle E_a(\cdot,s)u_j^\chi, u_j^\chi\rangle\ S\varphi(s_j^\chi)$$
$$= (\mathrm{hyp})_{a,s} + (\mathrm{ell})_{a,s} + \sum_{j=1}^{4} I_{a,j}(s,\chi,\varphi)$$

where:

(i) $$(\mathrm{hyp})_{a,s} = \sum_{\{\gamma\}_{\mathrm{hyp}}} \chi(\gamma)\left(\int_{\gamma_0} E_a(z,s)\right)H_s^c\ \varphi(a(\gamma))$$

(ii) $$(\mathrm{ell})_{a,s} = \sum_{\{\gamma\}_{\mathrm{ell}}} \chi(\gamma)\frac{1}{\nu(\gamma_0)}\ E_a(z_\gamma,s)L_s^c\ \varphi(k(\gamma))$$

(iii) $$I_{a,1}(s,\chi,\varphi) = \left[\sum_{\beta=1}^{h} 2\xi_{a\beta}(s,\chi)\right]\ \frac{\Gamma(1/2)\Gamma(s-1/2)}{\Gamma(s)}\ M\varphi(s/2)$$

where $$\xi_{a\beta}(s,\chi) = \sum_{\gamma\in\Gamma_\beta}{}' \chi(\gamma) \sum_{\substack{\tau\in\Gamma_\beta\backslash\Gamma/\Gamma \\ \tau^{-1}\gamma\,\tau\,\notin\,\Gamma_a}} \frac{1}{|c|^s} \qquad (\mathrm{Re}\,s > 1)$$

where: $\sigma_a^{-1}\tau^{-1}\gamma\tau\sigma_a = \begin{pmatrix} * & * \\ c & * \end{pmatrix}$, $\Sigma'$ omits $\gamma = \mathrm{id}$ and $\tau^{-1}\gamma\tau \notin \Gamma_a$ is vacuous if $a \neq \beta$.

Equivalently, $\xi_{a\beta}(s,\chi) = \varphi_{a\beta}(s,\chi)\left[\sum_{n\neq 0} \frac{e^{2\pi i n a(\beta)}}{|n|^s}\right]$ where: $\Gamma_\beta = \langle\gamma_{\beta 0}\rangle$ and $\chi(\gamma_{\beta 0}) = e^{2\pi i a(\beta)}$.

(iv) $$I_{a,2}(s,\chi,\varphi) = \frac{\xi^*(s)}{(4\pi)^{(s+1)/2}\Gamma[\frac{s+1}{2}]} \int_{-\infty}^{\infty} \frac{\Gamma(\frac{s}{2}+ir)\ \Gamma(\frac{s}{2}-ir)}{\Gamma(ir)\ \Gamma(-ir)}\ h(r)\ dr\ +$$
$$+\ 2\varphi_{aa}(\tfrac{s}{2}-1,\chi)\ \tilde{h}[\tfrac{is}{2}]$$

$$(\tilde{h}(\tfrac{is}{2}) = h(r);\ \rho^*(s) = \pi^{-s/2}\,\Gamma(\tfrac{s}{2})\rho(s)).$$

(v) $$I_{a,3}(s,\chi,\varphi) = \frac{\Gamma(\frac{s}{2})^2}{4\pi^s}\int_{-\infty}^{\infty} \Gamma(\tfrac{s}{2}+ir)\Gamma(\tfrac{s}{2}-ir)\Big[\sum_{\beta=1}^{h(\chi)} R_{\beta a}(\tfrac{1}{2}+ir,s,\chi)\Big]\, h(r)dr$$

(vi) $$I_{a,4}(s,\chi,\varphi) = \sum_{\substack{\text{poles } s_j \text{ of} \\ E_s \text{ in } (1/2,1)}} h(r_j)\ \Big\{\sum_{\beta,\gamma=1}^{h(\chi)} [\Phi^*(s_j,\bar{\chi})^{-1}]_{\beta\gamma} R_a[E^*_\beta(\cdot,s_j)E^*_\gamma(\cdot,s_j),\ s]\Big\}$$

<u>Proof</u> The spectral side is clear. On the other hand, $I_a(s,\chi,\varphi) = (\text{hyp}) + (\text{ell}) + (II_a)$, where (hyp) and (ell) are worked out as in § 2A, and where

$$II_a(s,\chi,\varphi) = \int_{\Gamma\backslash h} E_a(z,s)\Big[\sum_{\{\gamma\}\text{par}} K_{\{r\}}(z,z,\chi) + K_{\{e\}}(z,z,\chi) - (K_{eis} + K_{res} + \delta_0^\chi K_1)\Big]dz$$

($\delta_0^\chi = 1$ if $\chi \equiv 1$ on $\Gamma$, 0 otherwise). Let [*] denote the bracketed expression.

By the standard unfolding method, one has for Res > 1:

$$II_a(s,\chi,\varphi) = \int_0^\infty y^{s-2}\theta_a^0[*](y)\ dy$$

We now break up [*] into pieces whose $0^{th}$ fourier coefficients decay nicely in the $a^{th}$ cusp.

First consider $\sum_{\{\gamma\}_{par}} K_{\{\gamma\}}(z,z,\chi) = \sum_{\beta=1}^{h} \sum_{\{\gamma\}_{par,\beta}} K_{\{\gamma\}}(z,z,\chi)$, $\{\gamma\}_{par,\beta}$ running over the conjugacy classes of elements in $\Gamma_\beta$. Let us set:

(3.1) For $\beta \neq a$, $$K^*_{\beta,1}(z,z,\chi) = \sum_{\{\gamma\}_{par,\beta}} K_{\{\gamma\}}(z,z,\chi),$$

$$K_{a\beta,1}(y,\chi) = \theta_a^0[K^*_{\beta,1}(\cdot,\cdot,\chi)](y)$$

(3.1.1) $$K^*_{a,1}(z,z,\chi) = \sum_{n\neq 0} \chi(\gamma_{a0}^n) \sum_{\substack{\tau\in\Gamma_a\backslash\Gamma \\ [\tau]\neq id}} k(z,\tau^{-1}\gamma_{a0}^n\tau z),$$

$$K_{aa,1}(y,\chi) = \theta_a^0(K^*_{a,1})(y)$$

$$(3.1.2) \qquad K_{\alpha,1}(y,\chi) = \sum_{\beta=1}^{h} K_{\alpha\beta,1}(y,\chi).$$

Clearly, $\sum_{\{\gamma\}_{par}} K_{\{\gamma\}}(z,z,\chi) = \sum_{\beta=1}^{h} K^{*}_{\beta,1}(z,z,\chi) + \sum_{n\neq 0} \chi(\gamma^n_{\alpha 0})\ k(z,\gamma^n_{\alpha 0}z)$ and $\sum_{n\neq 0} \chi(\gamma^n_{\alpha 0})\ k(\sigma_\alpha z,\gamma^n_{\alpha 0}\sigma_\alpha z) = \sum_{n\neq 0} e^{2\pi i n a}\varphi[\frac{n^2}{y^2}]$. We isolate this term because it doesn't decay as $y \to \infty$. On the other hand $K_{\alpha,1}(y,\chi)$ does decay nicely, so we set:

$$(3.1.3) \qquad I_{\alpha,1}(s,\chi,\varphi) = \int_0^\infty y^{s-2} K_{\alpha,1}(y,\chi)\ dy.$$

It follows that $II_\alpha((s,\chi,\varphi)) = I_{\alpha,1} + III_\alpha$, where

$$(3.2) \qquad III_\alpha(s,\chi,\varphi) = \int_0^\infty y^{s-2}[\sum_n e^{2\pi i n a}\varphi[\frac{n^2}{y^2}] - \theta^0_\alpha(K_{eis} + K_{res} + \delta^\chi_0\ K_1)(y)]dy.$$

We can continue to break $III_\alpha$ up into large and small parts as $y^{(\alpha)} \to \infty$. We begin by recalling that $K_{eis}(z,z,\chi) = \sum_{\beta=1}^{h(\chi)} \frac{1}{4\pi} \int_{-\infty}^{\infty} |E_\beta(z,\tfrac{1}{2}+ir,\chi)|^2 h(r)\ dr.$
Let $A^0(z,s,\chi)$ be the matrix of constant terms: $A^0(z,s,\chi) = \{\theta^0_\alpha[E_\beta(\cdot,s,\chi)]\}$; i.e. $A^0_{\alpha\beta}(z,s,\chi) = \delta^\chi_{\alpha\beta} y^s_\alpha + \varphi_{\beta\alpha}(s,\chi)\ y^{1-s}_\alpha$ ($\delta^\chi_{\alpha\beta} = 1$ if $\alpha = \beta$ and $\chi \equiv 1$ on $\Gamma_\alpha$, 0 otherwise. $y_\alpha = \mathrm{Im}(\sigma_\alpha z)$).

Then $\theta^0_\alpha[|E_\beta(\cdot,s,\chi)|^2](y_\alpha) = |A^0_{\alpha\beta}(z,s,\chi)|^2 + \theta^0_\alpha|G_{\beta\alpha}(\cdot,s,\chi)|^2(y_\alpha)$. It is well-known (see § 1 and Appendix) that the second term decays rapidly as $y \to \infty$, so we isolate it and define:

$$(3.3) \qquad I_{\alpha\beta,3}(s,\chi,\varphi) = \frac{1}{4\pi} \int_0^\infty y^{s-2}[\int_{-\infty}^{\infty} \theta^0_\alpha|G_{\beta\alpha}(\cdot,\tfrac{1}{2}+ir,\chi)|^2(y)h(r)dr]\ dy,$$

$$I_{\alpha,3}(s,\chi,\varphi) = \sum_{\beta=1}^{h(\chi)} I_{\alpha\beta,3}(s,\chi,\varphi).$$

In a similar way, we may break up $\theta_\alpha^0 K_{res}$ into large and small parts. To do this, it is convenient to use the non-orthonormal basis $\{E^*(z,s_j,\chi)\}$, given by residues of Eisenstein series. We must digress momentarily to work out the inner products $\langle E_\alpha^*(z,s_j,\chi), E_\beta^*(z,s_j,\chi)\rangle$ (residues at different poles are of course orthogonal).

We claim $\langle E_\alpha^*(z,s_j,\chi), E_\beta^*(z,s_j,\chi)\rangle = \varphi_{\alpha\beta}^*(s_j,\chi)$ (cf. [Ro], p. 302).

Indeed, define as usual the cut-off Eisenstein series:

$$\tilde{E}_{\alpha,A}(z,s,\chi) = \begin{cases} E_\alpha(z,s,\chi) - (\delta_{\alpha\beta} y_\beta^s + \varphi_{\alpha\beta}(s)\, y_\beta^{1-s}) & z \in F_A^c(\beta) \\ E_\alpha(z,s,\chi) & z \in F_A \end{cases}$$

Also define the cut off residues: at a pole $s_*$ of $E_A(z,s,\chi)$, let

$$\tilde{E}^*_{\alpha,A}(z,s_*,\chi) = \begin{cases} E_\alpha^*(z,s_*,\chi) - \varphi_{\alpha\beta}^*(s_*)\, y_\beta^{1-s_*} & z \in F_A^c(\beta) \\ E_\alpha^*(z,s_*,\chi) & z \in F_A \end{cases}$$

Consider the inner product:

$$\int_F \tilde{E}^*_{\alpha,A}(z,s_*,\chi)\, \overline{\tilde{E}_{\beta,A}(z,s,\chi)}\, dz.$$

Following the standard argument [(C-S)], the integral over F breaks up into $\int_{F_A} E_\alpha^*(z,s_*,\chi)\, \overline{E_\beta}(z,s,\chi)\, dz + \sum_{\ell=1}^{h} \int_{y_\ell \geq A} \tilde{E}^*_{\alpha,A}\, \overline{\tilde{E}_{\beta,A}}\, dz$, where $F_A$ is the compact part of F and where $y_\ell \geq A$ are the ends $F_A^c(\ell)$ (A5).

Apply Green's identity: $\int_\Omega u\Delta v - v\Delta u = \int_{\partial\Omega} u\frac{\partial v}{\partial n} - v\frac{\partial y}{\partial n}$ to both integrals. The only boundary terms that do not cancel are those involving the zero$^{th}$ fourier coefficients integrated over the horocycles $y_\ell = A$. We thus get:

$$\int_F \tilde{E}^*_{\alpha,A}(z,s_*,\chi)\, \overline{\tilde{E}_{\beta,A}(z,s,\chi)}\, dz = \frac{1}{s_*(1-s_*)-\bar{s}(1-\bar{s})} \frac{1}{A} \sum_{\ell=1}^{h} [\varphi^*_{\alpha\ell}(s_*,\chi)A^{1-s_*}]$$

$$\times\ [\delta_{\beta\ell}A^{\bar{s}}\bar{s}+\overline{\varphi_{\beta\ell}(s,\chi)}A^{1-\bar{s}}(1-\bar{s})] - [(1-s_*)\varphi^*_{\alpha\ell}(s_*,\chi)A^{1-s_*}][\delta_{\beta\ell}A^{\bar{s}}+\overline{\varphi_{\beta\ell}(s,\chi)}\, A^{1-\bar{s}}]$$

$$= \frac{A^{\bar{s}-s_*}\varphi^*_{\alpha\beta}(s_*,\chi)}{\bar{s}-s_*} + \frac{A^{1-(s_*+\bar{s})}}{1-(s_*+\bar{s})} \sum_{\ell=1}^{h} \varphi^*_{\alpha\ell}(s_*,\chi)\overline{\varphi_{\beta\ell}(s,\chi)}.$$

Now we let $s \to s_*$. Clearly the first term has residue $\varphi^*_{\alpha\beta}(s_*,\chi)$ at the pole. The second term has residue $\frac{A^{1-2s_*}}{1-2s_*} \sum_{\ell=1}^{h} \varphi^*_{\alpha\ell}(s_*,\chi)\varphi_{\beta\ell}(s_*,\bar{\chi})$. This term drops out as $A \to \infty$ since $s_* > 1/2$. It follows that $\langle E^*_\alpha(s_*,\chi), E^*_\beta(\cdot,s_*,\chi)\rangle = \varphi^*_{\alpha\beta}(s_*,\chi)$, as desired. From this and the fact that $\varphi^*_{\alpha\beta}(s_*,\chi) = \varphi^*_{\beta\alpha}(s_*,\bar{\chi})$, we have:

$$K_{res}(z,w,\chi) = \sum_{\substack{\text{poles } s_j \text{ of} \\ E_s \text{ in } (1/2,1)}} h(r_j) \sum_{\beta,\gamma=1}^{h(\chi)} [\Phi^*(s_j,\bar{\chi})^{-1}]_{\beta\gamma}\, E^*_\beta(z,s_j,\chi)\, \bar{E}^*_\gamma(w,s_j,\chi).$$

It follows that $\theta^0_\alpha K_{res}(z,z,\chi)$ consists of the large part:

$$\sum_{\substack{\text{poles } s_j \text{ of} \\ E_s \text{ in } (1/2,1)}} h(r_j) \sum_{\beta,\gamma=1}^{h(\chi)} [\Phi^*(s_j,\bar{\chi})^{-1}]_{\beta\gamma}\, [\varphi^*_{\beta\alpha}(s_j,\chi)y_\alpha^{1-s_j}]\, [\bar{\varphi}^*_{\gamma\alpha}(s_j,\chi)y_\alpha^{1-s_j}]$$

$$= \sum_{\substack{\text{poles } s_j \text{ of} \\ E_s \text{ in } (1/2,1)}} \varphi^*_{\alpha\alpha}(s_j,\chi) y_\alpha^{2-2s_j} \text{ ; and the small part;}$$

$$\sum_{\substack{\text{poles } s_j \text{ of} \\ E_s \text{ in } (1/2,1)}} h(r_j) \sum_{\beta,\gamma=1}^{h(\chi)} [\Phi^*(s_j,\bar{\chi})^{-1}]_{\beta\gamma}\ \theta^0_\alpha[G^*_{\beta\alpha}(z,s_j,\chi)\ \bar{G}^*_{\gamma\alpha}(z,s_j,\chi)].$$

(3.4) Let us define $I_{\alpha,4}(s,\chi,\varphi)$ to be the Mellin transform (at s-1) of the small part, analogously to $I_{\alpha,3}$.

Summing up, we have: $III_\alpha(s,\chi,\varphi) = \sum_{j=2}^{4} I_{\alpha,j}(s,\chi,\varphi)$ where:

$$I_{\alpha,2}(s,\chi,\varphi) = \int_0^\infty y^{s-2}\Big[\sum_n e^{2\pi ina}\ \varphi[\tfrac{n^2}{y^2}] - \sum_{\beta=1}^{h(\chi)} \frac{1}{4\pi}\int_{-\infty}^{\infty} |A^0_{\alpha\beta}(y,\tfrac{1}{2}+ir,\chi)|^2\ h(r)dr$$
$$- \sum_{\text{poles } s_j} h(r_j)\ \varphi^*_{\alpha\alpha}(s_j,\chi)y^{2-2s_j} - \delta^\chi_0\ K_1\Big]\ dy,$$

$$I_{\alpha,3}(s,\chi,\varphi) = \int_0^\infty y^{s-2}\Big\{\sum_{\beta=1}^{h(\chi)} \frac{1}{4\pi}\int_{-\infty}^{\infty} drh(r)\,\theta^0_\alpha[|G_{\beta\alpha}(\cdot,\tfrac{1}{2}+ir,\chi)|^2]\Big\}\ dy,$$

$$I_{\alpha,4}(s,\chi,\varphi) = \int_0^\infty y^{s-2}\Big\{\sum_{\text{poles } s_j}\ \sum_{\beta,\gamma=1}^{h(\chi)} [\Phi^*(s_j,\bar{\chi})^{-1}]_{\beta\gamma}\theta^0_\alpha[G^*_{\beta\alpha}(z,s_j,\chi)\bar{G}^*_{\gamma\alpha}(z,s_j,\chi)]\Big\}dy$$

Our next task is to evaluate $I_{\alpha,j}(s,\chi,\varphi)$ in more familiar terms.

(1) $$I_{\alpha\alpha,1}(s) = \sum_{n\neq 0} e^{2\pi ina} \sum_{\substack{\tau\in\Gamma_\alpha\backslash\Gamma \\ [\tau]\neq id}} \int_0^\infty y^{s-2}\ \Big(\int_0^1 k[\sigma_\alpha(x+iy),\tau^{-1}\gamma^n_{\alpha 0}\tau\sigma_\alpha(x+iy)]\ dx\Big)\ dy.$$

The condition $[\tau] \neq$ id is unchanged under $\tau \longrightarrow \tau\gamma^n_{\alpha 0}$. Thus

$$I_{\alpha\alpha,1}(s) = \sum_{n\neq 0} e^{2\pi ina} \sum_{\substack{\tau\in\Gamma_\alpha\backslash\Gamma/\Gamma_\alpha \\ [\tau]\neq id}} \int_0^\infty y^{s-2}\ \Big(\int_{-\infty}^{\infty} k[\sigma_\alpha(x+iy),\tau^{-1}\gamma^n_{\alpha 0}\tau\sigma_\alpha(x+iy)]\ dx\Big)\ dy.$$

One has in general ([Za1]): If $\tau = \begin{pmatrix} a & b \\ c & d \end{pmatrix}$, $\tau \notin \Gamma_\infty$, then $\int_h k(z,\tau z) y^s dz = \frac{1}{|c|^s} V(s,t)$, where $t = \mathrm{tr}\,\tau$ and $V(s,t) = \int_h \varphi\left[\frac{|z^2+1-t^2/4|}{y^2}\right] y^s dz$ $\left[dz = \frac{dxdy}{y^2}\right]$. Here $\tau = \sigma_\alpha^{-1}\tau^{-1}\gamma_{\alpha 0}^n \tau\sigma_\alpha$, $t = 2$ and so

$$(3.5.1)\quad I_{\alpha\alpha,1}(s) = \sum_{n\neq 0} \chi(\gamma_{\alpha 0}^n)\Bigg[\sum_{\substack{\tau\in\Gamma_\alpha\backslash\Gamma/\Gamma_\alpha \\ [\tau]\neq id}} \frac{1}{|c_n|^s}\Bigg] V(s,2) = \varphi_{\alpha\alpha}(s,x)\ (s,2)$$

(in the $n^{th}$ term, $c_n = c(\sigma_\alpha^{-1}\tau^{-1}\gamma_{\alpha 0}^n\tau\sigma_\alpha)$). One has further (cf. [Za1]): $V(s,2) = 2\,\frac{\Gamma(\frac{1}{2})\Gamma(s-\frac{1}{2})}{\Gamma(s)}\,M\varphi(s/2)$. Thus $I_{\alpha\alpha,1}(s) = 2\varphi_{\alpha\alpha}(s,\chi)\frac{\Gamma(\frac{1}{2})\Gamma(s-\frac{1}{2})}{\Gamma(s)} \times M\varphi(s/2)$.

Similarly we have, for $\alpha \neq \beta$:

$$\begin{aligned}(3.5.2)\quad I_{\alpha\beta,1}(s) &= \sum_{n\neq 0}\chi(\gamma_{\beta 0}^n)\sum_{\tau\in\Gamma_\beta\backslash\Gamma}\int_0^\infty y^{s-2}\Big\{\int_0^1 k[\sigma_\alpha(x+iy),\tau^{-1}\gamma_{\beta 0}^n\tau\sigma_\alpha(x+iy)]\ dx\Big\}\ dy \\ &= \sum_{n\neq 0}\chi(\gamma_{\beta 0}^n)\sum_{\tau\in\Gamma_\beta\backslash\Gamma/\Gamma_\alpha}\int_0^\infty y^{s-2}\Big\{\int_{-\infty}^\infty k[\sigma_\alpha(x+iy),\tau^{-1}\gamma_{\beta 0}^n\tau\sigma_\alpha(x+iy)]dx\Big\}dy \\ &= \Bigg[\sum_{n\neq 0}\chi(\gamma_{\beta 0}^n)\sum_{\tau\in\Gamma_\beta\backslash\Gamma/\Gamma_\alpha}\frac{1}{|c_n|^s}\Bigg] V(s,2) \\ &= 2\varphi_{\alpha\beta}(s,\chi)\frac{\Gamma(\frac{1}{2})\Gamma(s-\frac{1}{2})}{\Gamma(s)}\,M\varphi(s/2) \qquad c_n = c(\sigma_\alpha^{-1}\tau^{-1}\gamma_{\alpha 0}^n\tau\sigma_\alpha).\end{aligned}$$

These series converge absolutely for $\mathrm{Re}\,s > 1$, as do all the series which appear in the computation.

We note that $I_{\alpha\beta,1}(s,\chi,\varphi) = \int_{\Gamma\backslash h} E_\alpha(z,s)\,K_{\{\gamma\}_{par,\beta}}(z,z,\chi)\,dz$ for $\beta \neq \alpha$. So these $I_{\alpha\beta}$ have analytic continuations to $\mathbb{C}$ with poles among the poles of $E_\alpha(z,s)$. The $I_{\alpha\beta,1}(s)$ may be quickly related to the $\varphi_{\alpha\beta}(s)$; we pause to do this.

We have (with $\chi(\gamma_{\beta 0}) = e^{2\pi i a(\beta)}$):

$$(3.5.3) \qquad (\beta \neq \alpha): I_{\alpha\beta,1}(s,\chi,\varphi)$$
$$= \sum_{n\neq 0} \chi(\gamma_{\beta 0}^{n}) \int_{\Gamma\backslash h} E_{\alpha}(z,s) \, [\sum_{\tau\in\Gamma_{\beta}\backslash\Gamma} k(z,\tau^{-1}\gamma_{\beta 0}^{n}\tau z)] \, dz$$
$$= \sum_{n\neq 0} \chi(\gamma_{\beta 0}^{n}) \int_{\Gamma_{\beta}\backslash h} E_{\alpha}(z,s) \, k(z,\gamma_{\beta 0}^{n} z) \, dz$$
$$= \sum_{n\neq 0} e^{2\pi i n a(\beta)} \int_{\Gamma_{\infty}\backslash h} E_{\alpha}(\sigma_{\beta} z,s) \, k(\sigma_{\beta} z,\gamma_{\beta 0}^{n}\sigma_{\beta} z) \, dz$$
$$= \sum_{n\neq 0} e^{2\pi i n a(\beta)} \int_{0}^{\infty} \varphi[\frac{n^2}{y^2}] \, [\delta_{\alpha\beta} y^{s} + \varphi_{\alpha\beta}(s) \, y^{1-s}] \, \frac{dy}{y^{s}}$$

There are no convergence problems as $y \to 0$ if $\varphi \in C_0^{\infty}(\mathbb{R})$ and none as $y \to \infty$ if $\alpha \neq \beta$. We thus have for $\alpha \neq \beta$:

$$(3.5.4) \qquad I_{\alpha\beta,1}(s,\chi,\varphi) = \varphi_{\alpha\beta}(s) \, [\sum_{n\neq 0} \frac{e^{2\pi i n a(\beta)}}{|n|^{s}}] \, M\varphi(s/2) \qquad (\mathrm{Re}\, s > 1)$$

The case $\alpha = \beta$ is slightly different since $K^{*}_{\alpha,1}(z,z,\chi)$ is not $\Gamma$-invariant. However, $I_{\alpha\alpha}(s,\chi,\varphi)$ is a sum of terms of the form

$$(3.5.5) \qquad \sum_{\substack{\tau\in\Gamma_{\alpha}\backslash\Gamma/\Gamma_{\alpha} \\ [\tau]\neq id}} \int_{h} y^{s} \, k[\sigma_{\alpha} z,\tau^{-1}\gamma_{\alpha 0}^{m}\tau\sigma_{\alpha} z) \, dz$$
$$= \sum_{\substack{\tau\in\Gamma_{\alpha}\backslash\Gamma/\Gamma_{\alpha} \\ [\tau]\neq id}} \int_{h} \mathrm{Im}(\sigma_{\alpha}^{-1}\tau^{-1} z)^{s} \, k(z,\gamma_{\alpha 0}^{m} z) \, dz$$
$$= \sum_{n=-\infty}^{\infty} \sum_{\substack{\tau\in\Gamma_{\alpha}\backslash\Gamma/\Gamma_{\alpha} \\ [\tau]\neq id}} \int_{\gamma_{0\alpha}^{n} F_{\alpha}} \mathrm{Im}(\sigma_{\alpha}^{-1}\tau^{-1} z)^{s} \, k(z,\gamma_{\alpha 0}^{m} z) \, dz,$$

(where $F_{\alpha}$ is a fundamental domain for $\Gamma_{\alpha}$),

$$= \sum_{n=-\infty}^{\infty} \sum_{\substack{\tau\in\Gamma_a\backslash\Gamma/\Gamma_a \\ [\tau]\neq id}} \int_{F_a} \mathrm{Im}(\sigma_a^{-1}\tau^{-1}\gamma_{a0}^{n}z)^s\, k(z,\gamma_{a0}^{m}z)\, dz$$

$$= \sum_{\substack{\tau\in\Gamma/\Gamma_a \\ [\tau]\neq id}} \int_{F_a} \mathrm{Im}(\sigma_a^{-1}\tau^{-1}z)^s\, k(z,\gamma_{a0}^{m}z)\, dz$$

$$= \int_{F_a} [E_a(z,s) - \mathrm{Im}(\sigma_a^{-1}z)^s]\, k(z,\gamma_{a0}^{m}z)\, dz$$

$$= \varphi_{aa}(s)\, n^{-s} \mathbf{M}\varphi(s/2)$$

So $I_{aa,1}(s,\chi,\varphi) = \varphi_{aa}(s)\, \mathbf{M}\varphi(s/2) \sum_{n\neq 0} \frac{e^{2\pi ina}}{|n|^s}$.

(2) We turn to $I_{a,2}(s,\chi,\varphi)$. Note that the shape of $I_{a,2}$ depends on whether $\chi \equiv 1$ on $\Gamma_a$ or not. If $\chi \not\equiv 1$ on $\Gamma_a$, then

$$(3.6) \qquad I_{a,2}(s,\chi,\varphi) = \int_0^\infty y^{s-2} \left[\sum_n e^{2\pi ina} \varphi\left[\frac{n^2}{y^2}\right]\right] dy$$

($\chi(\gamma_{a0}) = e^{2\pi ia}$, $0 < a < 1$).

By Poisson summation,

$$\sum_n e^{2\pi ina} \varphi\left[\frac{n^2}{y^2}\right] = y \sum_n \psi[y(a+n)]$$

where $\psi(y) = \int_{-\infty}^{\infty} \varphi(u^2)\, e^{2\pi iny} du$. So

$$I_{a,2}(s,\chi,\varphi) = \left[\sum_n \frac{1}{|a+n|^s}\right] \mathbf{M}\psi(s) = [\zeta(s,a) + \zeta(s,1-a)]\, \mathbf{M}\psi(s)$$

In [Za1] it is shown that

$$\mathbf{M}\psi(s) = \frac{2^{-(s-2)}\Gamma(s/2)}{\pi^{s+3/2}\Gamma\left[\frac{s+1}{2}\right]} \int_{-\infty}^{\infty} \Gamma(s/2 - ir)\Gamma(s/2 - ir)\, r(\mathrm{sh}\, \pi r) h(r)\, dr,$$

giving our final expression for $I_{a,2}$ (sh = hyperbolic sine).

If $\chi \equiv 1$ on $\Gamma_a$, then

(3.7)
$$I_{a,2}(s,\chi,\varphi)$$
$$= \int_0^\infty y^{s-2}\Big[\sum_n \varphi\Big[\frac{n^2}{y^2}\Big]\Big] - \sum_{\beta=1}^{h(\chi)} \frac{1}{4\pi}\int_{-\infty}^{\infty} \big| \delta_{a\beta} y^{1/2+ir} + \varphi_{\beta a}(\tfrac{1}{2}+ir,\chi)\, y^{1/2+ir}\big|^2$$
$$\times\, h(r)\, dr - \sum_{\text{poles } s_j} h(r_j)\, \overset{*}{\varphi}_{aa}(s_j,\chi) y^{2-2s_j} - \delta_0^\chi K_1\Big]\, dy,$$
$$= \int_0^\infty y^{s-2}\Big[\sum_n \varphi\Big[\frac{n^2}{y^2}\Big] - \frac{1}{4\pi}\int_{-\infty}^{\infty} h(r)\, dr\Big]\, dy$$
$$- \int_0^\infty y^{s-2}\Big\{\Big[\frac{1}{4\pi}\int_{-\infty}^{\infty} 2\mathrm{Re}\varphi_{aa}(\tfrac{1}{2}+ir,\chi)\, y^{1-2ir}\, h(r)\, dr\Big]$$
$$- \sum_{\text{poles } s_j} h(r_j)\, \overset{*}{\varphi}_{aa}(s_j,\chi) y^{2-2s_j} - \delta_0^\chi K_1\Big\}\, dy.$$

The first term is computed in [Za1], and clearly has nothing to do with $\Gamma$. It equals:

$$\frac{\overset{*}{\zeta}(s)}{(4\pi)^{\frac{s+1}{2}}\,\Gamma\big[\frac{s+1}{2}\big]} \int_{-\infty}^{\infty} \frac{\Gamma(\frac{s}{2}+ir)\Gamma(\frac{s}{2}-ir)}{\Gamma(ir)\Gamma(-ir)}\, h(r)\, dr \qquad (\mathrm{Re}s > 1).$$

The second term may be computed by shifting the contour in the inner integral. We may assume $h(r) = h(-r)$; and then:

$$\int_{-\infty}^{\infty} 2\mathrm{Re}\varphi_{aa}(\tfrac{1}{2}+ir,\chi)\, y^{1-2ir}\, h(r)\, dr = 2\int_{\mathrm{Re}\tau=1/2} \varphi_{aa}(\tau,\chi)\, y^{2-2\tau}\, \tilde{h}\Big[\frac{i\tau}{2}\Big]\, \frac{d\tau}{i}.$$

($\tilde{h}\big[\frac{ir}{2}\big] = h(r)$ is analytic by our assumption on $\varphi$). Shifting to $\mathrm{Re}\tau > 1$ cancels out the residual terms, and yields the following expression for the second term:

$$(3.7.1)\qquad 2\int_{\operatorname{Re}\tau=s_0} \varphi_{aa}(\tau,\chi)\, y^{2-2\tau}\, \tilde{h}[\tfrac{i\tau}{2}]\, \frac{d\tau}{i} \qquad (\text{for any } s_0>1).$$

Setting $\operatorname{Re}\tau = \frac{1}{2}\operatorname{Re}s$, the second term may be evaluated by Mellin inversion as:

$$(3.7.2)\qquad 2\varphi_{aa}[\tfrac{s}{2}-1,\chi]\ \tilde{h}[\tfrac{is}{2}].$$

We conclude:

$$(3.7.3)\qquad I_{a,2}(s,\chi,\varphi) = \frac{\zeta^*(s)}{(4\pi)^{\frac{s+1}{2}}\ \Gamma[\frac{s+1}{2}]} \int_{-\infty}^{\infty} \frac{\Gamma(\frac{s}{2}+ir)\Gamma(\frac{s}{2}-ir)}{\Gamma(ir)\Gamma(-ir)}\, h(r)\, dr + 2\varphi_{aa}[\tfrac{s}{2}-1,\chi]\ \tilde{h}[\tfrac{is}{2}]$$

$(\operatorname{Re}s > 2)$.

It is clear that $I_{a,2}$ can be analytically continued.

$$(3)\quad I_{a\beta,3}(s,\chi,\varphi) = \frac{1}{4\pi}\int_0^{\infty} y^{s-2}\Big[\int_{-\infty}^{\infty} \theta_a^0 |G_{\beta a}(\cdot,\tfrac{1}{2}+ir,\chi)|^2\, h(r)\, dr\Big]\, dy,$$

so $I_{a\beta,3}$ is a kind of Rankin-Selberg convolution square of $E_\beta$. In the notation of § 1,

$$I_{a\beta,3}(s,\chi,\varphi) = \frac{\Gamma(s/2)^2}{4\pi^s}\int_{-\infty}^{\infty} \Gamma(\tfrac{1}{2}+ir)\ \Gamma(\tfrac{1}{2}-ir)\ R_{a\beta}(\tfrac{1}{2}+ir,s,\chi)\ h(r)\ dr.$$

Our problem here is to show that $R_{a\beta}(\frac{1}{2}+ir,s,\chi)$ can be analytically continued to $\mathbb{C}$ from $\operatorname{Re}s > 1$, or equivalently that $\int_0^{\infty} y^{s-2}\theta_a^0|G_{\beta a}(\cdot,\tau,\chi)|^2\, dy$ can be.

The argument for this essentially our original truncation approach to the trace formula and also essentially appears in [Za2]. However, both because [Za2] only explicitly treats $\Gamma = SL_2(\mathbb{Z})$ and also for the sake of completeness, we will give the general argument here. To facilitate comparison with [Za2] we will invoke the notation $R_\alpha[|E_\beta(\cdot,\tau)|^2, s]$ for $\int_0^\infty y^{s-2}\theta_\alpha^0|G_{\beta a}(\cdot,\tau,\chi)|^2\,dy$.

Let $F_T$ denote the compact part of the fundamental domain (as in A5). Consider the integral $\int_{F_T} E_\alpha(z,s)\ |E_\beta(z,\tau,\chi)|^2\,dz$ (which naturally appears in the truncation approach to Tr $E_s R_\varphi$). Let $F_T = \bigcup_{\gamma\in\Gamma} \gamma F_T$. Then

$$F_T = \bigcup_{\gamma\in\Gamma} \gamma[F - \bigcup_{\delta=1}^{h} (F_T^c(\delta))] = h - \bigcup_{\delta=1}^{h} \bigcup_{\sigma\in\Gamma\backslash\Gamma_\delta} \bigcup_n \sigma\gamma_{0\delta}^n ((F_T^c(\delta))$$

$$= h - \bigcup_{\delta=1}^{h} \bigcup_{\sigma\in\Gamma\backslash\Gamma_\delta} \sigma(y_\delta \geq T) \qquad \text{(for T large enough)}$$

$$= h_{\alpha,T} - \bigcup_{\substack{\sigma\in\Gamma\backslash\Gamma_\alpha \\ \sigma\in\Gamma_\alpha}} \sigma(y_\alpha \geq T) - \bigcup_{\delta\neq\alpha} \bigcup_{\sigma\in\Gamma\backslash\Gamma_\delta} \sigma(y_\delta \geq T)$$

where $h_{\alpha,T} = \{z\colon y^\alpha \leq T\}$.

It follows that

$$(3.8) \qquad \int_{F_T} E_\alpha(z,s)\ |E_\beta(z,\tau,\chi)|^2\,dz = \int_{\Gamma_\alpha\backslash F_T} (y_\alpha)^s\ |E_\beta(z,\tau,\chi)|^2\,dz$$

$$= \int_{\Gamma_\alpha\backslash h_{\alpha,T}} \text{(same)} - \sum_{\substack{\sigma\in\Gamma_\gamma\backslash\Gamma/\Gamma_\alpha \\ \sigma\notin\Gamma_\alpha}} \int_{\sigma(y_\alpha\geq T)} \text{(same)}$$

$$- \sum_{\delta\neq\alpha} \sum_{\sigma\in\Gamma_\alpha\backslash\Gamma/\Gamma_\alpha} \int_{\sigma(y_\gamma\geq T)} \text{(same)}$$

((same) being $\Gamma_\alpha$-invariant)

$$(3.8.1) \qquad = \int_0^T y^{s-2}\theta_\alpha^0|E_\beta(\cdot,\tau,\chi)|^2\,dy$$

$$- \sum_{\substack{\sigma\in\Gamma_\alpha\backslash\Gamma/\Gamma_\alpha \\ \sigma\notin\Gamma_\alpha}} \int_{y_\alpha\geq T} [y_\alpha(\sigma z)]^s |E_\beta(z,\tau,\chi)|^2 \, dz$$

$$- \sum_{\delta\neq\alpha} \sum_{\sigma\in\Gamma_\alpha\backslash\Gamma/\Gamma_\delta} \int_{y_\delta\geq T} [y_\alpha(\sigma z)]^s |E_\beta(z,\tau,\chi)|^2 \, dz$$

The second term is $\displaystyle\sum_{\sigma\in\Gamma_\alpha\backslash\Gamma/\Gamma_\alpha} \sum_n \int_{F_T^c(\alpha)} [y_\alpha(\sigma\gamma_{\alpha_0}^n z)]^s |E_\beta(z,\tau,\chi)|^2 \, dz$

$$= \sum_{\substack{\sigma\in\Gamma_\alpha\backslash\Gamma \\ \sigma\notin\Gamma_\alpha}} \int_{(y_\alpha\geq T)\cap F} [y_\alpha(\sigma z)]^s |E_\beta(z,\tau,\chi)|^2 \, dz$$

$$= \int_{(y_\alpha\geq T)\cap F} [E_\alpha(z,s) - (y_\alpha)^s] |E_\beta(z,\tau,\chi)|^2 \, dz.$$

The same argument with $\delta$ replacing $\alpha$ evaluates the third term as $\displaystyle\sum_{\delta\neq\alpha} \int_{F_T^c(\alpha)} E_\alpha(z,s)|E_\beta(z,\tau,\chi)|^2 \, dz$, which converges nicely if $\mathrm{Re}\, s > 1$.

Summing up:

$$(3.8.2) \quad \int_{F_T} E_\alpha(z,s) \, |E_\beta(z,\tau)|^2 \, dz = \int_0^T y^{s-2} \theta_\alpha^0 |E_\beta(\cdot,\tau,\chi)|^2 \, dy$$
$$- \int_{F_\alpha^c(T)} [E_\alpha(z,s) - (y^\alpha)^s] |E_\beta(z,\tau)|^2 \, dz$$
$$- \sum_{\delta\neq\alpha} \int_{F_\delta^c(T)} E_\alpha(z,s)|E_\beta(z,\tau,\chi)|^2 \, dz$$

We now rewrite the first term:

$$\int_0^T y^{s-2}\theta_\alpha^0(|E_\beta|^2) \, dy = \int_0^T y^{s-2}\theta_\alpha^0(|A_{\beta\alpha}|^2) \, dy + \int_0^T y^{s-2}\theta_\alpha^0(|G_{\beta\alpha}|^2) \, dy.$$

Let $h_{\beta\alpha T}(s) = \int_0^T y^{s-2}(|A_{\beta\alpha}|^2)\, dy$. Then:

$$\int_0^T y^{s-2}\theta_\alpha^0(|E_\beta|^2)\, dy = h_{\beta\alpha T}(s) + R_\alpha(|E_\beta|^2,s) - \int_T^\infty y^{s-2}\theta_\alpha^0(|G_{\beta\alpha}|^2)\, dy.$$

Next we rewrite the second term:

$$\int_{F_\alpha^c(T)} [E_\alpha(z,s)-(y_\alpha)]^s|E_\beta|^2\, dz = \int_{(y_\alpha\geq T)\cap F} [E_\alpha(z,s) - A_{\alpha\alpha}(y,s)]|E_\beta|^2\, dz$$
$$+ \varphi_{\alpha\alpha}(s)\int_T^\infty y^{-1-s}|A_{\beta\alpha}(y,\tau,\chi)|^2\, dy + \varphi_{\alpha\alpha}(s)\int_T^\infty y^{-1-s}\,\theta_\alpha^0(|G_{\beta\alpha}|^2)\, dy.$$

(3.8.3) Thus: $\int_0^T y^{s-2}\theta_\alpha^0(|E_\beta|^2)\, dy - \int_{(y_\alpha\geq T)\cap F} [E_\alpha(z,s)-(y_\alpha)^s]|E_\beta|^2\, dz$

$$= h_{\beta\alpha T}(s) + R_\alpha(|E_\beta|^2,s) - \int_{(y_\alpha\geq T)\cap F} [E_\alpha(z,s) - A_{\alpha\alpha}(y,s)]|E_\beta|^2\, dz$$
$$- \int_T^\infty A_{\alpha\alpha}(y,s)\ \theta_\alpha^0(|G_{\beta\alpha}|^2)\ (y)\ \frac{dy}{y^2} - \varphi_{\alpha\alpha}(s)h_{\beta\alpha T}(1-s)$$
$$= h_{\beta\alpha T}(s) - \varphi_{\alpha\alpha}(s)h_{\beta\alpha T}(1-s) + R_\alpha(|E_\beta|^2,s)$$
$$- \int_{(y_\alpha\geq T)\cap F} [E_\alpha(z,s)|E_\beta|^2 - A_{\alpha\alpha}(y_\alpha,s)\ |A_{\beta\alpha}(y_\alpha,\tau,\chi)|^2]\, dz$$

It is clear that $E_\alpha(z,s)|E_\beta(z,\tau,\chi)|^2 - A_{\alpha\alpha}(y_\alpha,s)\ |A_{\beta\alpha}(y_\alpha,\tau,\chi)|^2$ is rapidly decaying in $(y_\alpha \geq T) \cap F$, so the first two terms of (3.8.3) combine to form a memorphic function if and only if $R_\alpha(|E_\beta|^2,s)$ does.

This leaves the third term of (3.8.2). As before, we re-write:

(3.8.4) $\int_{y_\delta\geq T} E_\alpha(z,s)\ |E_\beta|^2\, dz$

$$= \int_{y_\delta \geq T} [E_\alpha(z,s) - A_{\alpha\delta}(y_\delta,s)(|E_\beta|^2) \, dz + \varphi_{\alpha\delta}(s) \int_T^\infty y^{-1-s} \, \theta_\alpha^0(|E_\beta|^2)(y) \, dy$$

$$= \varphi_{\alpha\alpha}(s) h_{\beta\alpha T}(1-s) + \int_{y_\delta \geq T} E_\alpha(z,s) \, |E_\beta|^2 - A_{\alpha\delta}(y_\delta,s) |A_{\beta\alpha}(y_\delta,\tau,\chi)|^2 dz$$

Summing everything up, we have:

$$(3.8.5) \quad R_\alpha[|E_\beta(\cdot,\tau,\chi)|^2, s]$$
$$= \int_{F_T} E_\alpha(z,s) \, |E_\beta(z,\tau,\chi)|^2 \, dz - h_{\beta\alpha T}(s) - \sum_\delta \varphi_{\alpha\delta}(s) \, h_{\beta\delta T}(1-s)$$
$$+ \sum_\delta \int_{F_\alpha^c(T)} [E_\alpha(z,s) \, |E_\beta(z,\tau,\chi)|^2 - A_{\alpha\delta}(y_\delta,s) |A_{\beta\alpha}(y_\delta,\tau,\chi)|^2] \, dz$$

This formula gives the desired memorphic continuation of $R_{\alpha\beta}(\tau,s)$ to $\mathbb{C}$, and reduces all of its analytic properties to those of $E_\alpha(z,s)$.

(4) Our last object in this part is to confirm that $I_{\alpha,4}(s,\chi,\varphi)$ has a memorphic continuation to $\mathbb{C}$ with analytic properties deduced from those of $E_\alpha$. This will clearly follow from like statements about $R[E_\beta^*(\cdot,s_j) \, E_\gamma^*(\cdot,s_j), \, s]$. For this, we consider, as in step (3), the integrals:

$$\int_{F_T} E_\alpha(z,s) \, E_\beta^*(z,\tau,\chi) \, E_\gamma^*(z,\tau,\chi) \, dz$$

where $\tau$ is a simultaneous pole of $E_\beta(z,\tau,\chi)$ and $E_\gamma(z,\tau,\chi)$. The result is:

$$(3.9) \quad R_\alpha[E_\beta^*(\cdot,s_j) \, E_\gamma^*(\cdot,s_j), \, s]$$
$$= \int_{F_T} E_\alpha(z,s) \, E_\beta^*(z,\tau,\chi) \, E_\gamma^*(z,\tau,\chi) \, dz - h_{\beta\gamma\alpha T}^*(s) - \sum_\delta \varphi_{\alpha\delta}(s) \, h_{\beta\gamma\delta T}^*(1-s)$$

$$+ \sum_{\delta} \int_{F_\delta^c(T)} [E_a(z,s)\ [E_\beta^*(z,\tau,\chi)E_\gamma^*(z,\tau,\chi)] - A_{a\delta}(A_{\beta\delta}^* A_{\gamma\delta}^*)(y_\delta,s)]\ dz,$$

where $h_{\beta\gamma a T}^*(s) = \int_0^T y^{s-2}[\varphi_{\beta a}^*(\tau)\ \varphi_{\gamma a}^*(\tau)\ y^{2-2\tau}]\ dy.$ ∎

We next consider the higher weight case

(B) $\sigma = X_+E_s$, $\varphi \in S_{2,0}$.

Set: $I_a^+(s,\chi,\varphi) = \frac{1}{2i} Tr_\chi \pi_0[E^+E_a(g,s)\ R_\varphi^\Gamma - E^-E_a(g,s)\ R_{\varphi^*}^\Gamma]\ \pi_0$

$$= \frac{s}{2i}\int_{\Gamma\backslash G} [E_a(x,s)\,{}_{(2)}K_0(x,x,\chi) - E_a(x,s)\,{}_{(-2)}K_0^*(x,x,\chi)]\ dx.$$

<u>Theorem 3(B)</u>

$$I_a^+(s,\chi,\varphi) = \sum_{j=1}^{\infty} \langle Op(X_+E_{a,s})u_j^\chi, u_j^\chi\rangle\ S_2\varphi(S_j^\chi) = (hyp)_{a,s}^+ + I_{a,3}^+(s) + I_{a,3}^+(s)$$

where:

(i) $(hyp)_{a,s}^+ = \sum_{\{\gamma\}} \chi(\gamma)\ (\int_{\gamma_0} X_+E_{a,s})\ H_{s,+}^c\ \varphi(\delta_\gamma)$

(ii) $I_{a,3}^+(s) = \sum_{\beta=1}^{h(\chi)} I_{a\beta,3}^+(s,\chi,\varphi)$

where $I_{a\beta,3}^+(s,\chi,\varphi) = \frac{1}{4\pi}\int_{-\infty}^{\infty} R_{a\beta}(\frac{1}{2}+ir,s,\chi)\ A^+(\frac{1}{2}+ir,s)\ S\varphi(\frac{1}{2}+ir)\ dr$, and where $A^+$ is essentially a sum of products of Γ-values (to be given below);

(iii) $I_{a,4}^+(s)$

$$= \sum_{\substack{\text{poles } s_j \\ \text{of } E_{a,s}}} A^+(s_j,s)S_2\varphi(s_j)\ \{\sum_{\beta,\delta=1}^{h(\chi)} [\Phi^*(s_j,\bar{\chi})^{-1}]_{\beta\delta}\ R_{a,\beta\delta}^*(s_j,s,\chi)\}.$$

<u>Proof</u> The cuspidal side and the hyperbolic and elliptic terms of the right side work out exactly as in the cuspidal case.

The main thing is thus to work out:

$$(3.10)\quad II^{+}_{\alpha\beta,3}(s,\chi,\varphi) = \frac{s}{2i}\Big[[\int_{\Gamma\backslash G}[E_{\alpha}(x,s)_{(2)}[\varphi(e) + \sum_{\{\gamma\}_{par}} \chi(\gamma)\varphi(x^{-1}\gamma x)$$
$$- K_{eis}(x,x,\chi) - K_{res}(x,x,\chi)]\, dx$$
$$- \int_{\Gamma\backslash G} E_{\alpha}(x,s)_{(-2)}[\varphi(e) + \sum_{\{\gamma\}_{par}} \chi(\gamma)\varphi(x^{-1}\gamma x) - K^{*}_{eis} - K^{*}_{res}]\, dx$$

where:

(a) $$K_{eis}(x,y,\chi) = \frac{1}{4\pi}\int_{-\infty}^{\infty} E(x,\tfrac{1}{2}+ir,\chi)_{(-2)}\; \bar{E}(y,\tfrac{1}{2}+ir,\chi)\; S_{2}\varphi(\tfrac{1}{2}+ir)\, dr$$

$$K^{*}_{eis}(x,y,\chi) = \frac{1}{4\pi}\int_{-\infty}^{\infty} E(x,\tfrac{1}{2}+ir,\chi)_{(2)}\; \bar{E}(y,\tfrac{1}{2}+ir,\chi)\; S_{2}\varphi(\tfrac{1}{2}+ir)\, dr$$

(continuous spectral parts of $R_{\varphi}$)

(b) $K_{res}(x,y,\chi)$

$$= \sum_{\substack{\text{poles } s_j \\ \text{of } E_{\alpha,s}}} S_{2}\varphi(s_j)\; \{\sum_{\beta,\delta=1}^{h(\chi)} [\Phi^{*}(s_j,\bar{\chi})^{-1}]_{\beta\delta}\; E^{*}_{\beta}(x,s_j,\chi)_{(-2)}\; \bar{E}^{*}_{\delta}(y,s_j,\chi)\}$$

and similarly for $K^{*}_{res}$.

To see (3.10) (a), (b) we note that

$$R^{\Gamma}_{\varphi}E(x,\tfrac{1}{2}+ir,\chi) = S_{2}\varphi(\tfrac{1}{2}+ir)\; E(x,\tfrac{1}{2}+ir,\chi)_{(-2)}$$
$$R^{\Gamma}_{\varphi}E^{*}(x.s_j,x) = S_{2}\varphi(s_j)\; E^{*}(x,s_j,\chi)_{(-2)}, \qquad \text{and } R_{\varphi}1 = 0.$$

Following the weight 0 case, we unfold the integrals for $\mathrm{Re}\, s > 1$, and get:

$$II^{+}_{\alpha}(s,\chi,\varphi) = \frac{s}{2i}\{\int_{\Gamma_{\alpha}\backslash G} y_{\alpha}(x,s)_{(2)}\,\theta^{0}_{\alpha}[\varphi(e) + \sum_{\{\gamma\}_{par}} \chi(\gamma)\varphi(x^{-1}\gamma x)$$
$$- K_{eis}(x,x,\chi) - K_{res}(x,x,\chi)]\, dx - \int_{\Gamma_{\alpha}\backslash G} y_{\alpha}(x,s)_{(-2)}(\text{same}^{*})\}$$

$$= \frac{s}{2i} \int_0^\infty y^{s-2} \mathrm{Fm}(\theta_a^0(\text{same})(y)\ dy$$

where Fm is with respect to $^*$. Following the pattern in (3A), we get:

(3.11) $\beta \neq a$:
$$K^{+*}_{\beta,1}(z,z,\chi) = \mathrm{Fm} \sum_{\{\gamma\}_{\mathrm{par},\beta}} K_{\{\gamma\}}(z,z,\chi)$$
$$K^{+}_{a\beta,1}(y,\chi) = \theta_a^0[K^{+*}_{\beta,1}(\cdot,\cdot,\chi)(y)$$
$$K^{+*}_{a,1}(z,z,\chi) = \mathrm{Fm} \sum_{n\neq 0} \chi(\gamma_{a0}^n)[\sum_{\substack{\tau\in\Gamma_a\backslash\Gamma \\ [\tau]\neq \mathrm{id}}} k(z,\tau^{-1}\gamma_{a0}^n\tau z)]$$
$$K^{+}_{aa,1}(y,\chi) = \theta_a^0\ K^{+*}_{a,1}(y)$$

and

(3.11.1)
$$K^{+}_{a,1}(y,\chi) = \sum_{\beta=1}^{h} K_{a\beta,1}(y,\chi).$$

Here we use the notation $k(z,\sigma w) = \varphi(p_z^{-1}\sigma p_w)$ where $p_z \in P = AN$ is such that $p_z \cdot i = z$. k is not a point-pair invariant but is clearly the higher weight analogue.

We further set:

(3.11.2)
$$I^{+}_{a,1}(s,\chi,\varphi) = \int_0^\infty y^{s-2} K^{+}_{a,1}(y,\chi)\ dy,$$
so that $II^{+}_{a}(s,\chi,\varphi) = I^{+}_{a,1} + III^{+}_{a,1}$, where
$$II^{+}_{a,1}(s,\chi,\varphi) = S\int_0^\infty y^{s-2}\theta_a^0\ \{\mathrm{Im}[\sum_n \chi(\gamma_{a0}^n)k(z,\gamma_{a0}^n z) - K_{\mathrm{eis}} - K_{\mathrm{res}}]\}(y)\ dy.$$

Still following (3A), we note that

$$\theta_a^0\ [E(\tfrac{1}{2}+ir,\chi)_{(-2)}\ E(\tfrac{1}{2}+ir,\chi)]\ (y)$$
$$= \sum_{\beta=1}^{h(\chi)} [\delta_{a\beta}y^{1/2+ir} + y^{1/2-ir}\varphi_{\beta a}(\tfrac{1}{2}+ir,\chi)_{(-2)}]\,\overline{[\delta_{a\beta}y^{1/2+ir} + y^{1/2-ir}\varphi_{\beta a}(\tfrac{1}{2}+ir,\chi)]}$$
$$+ \theta_a^0[\sum_{\beta=1}^{h(\chi)} G_{\beta a}(z,\tfrac{1}{2}+ir,\chi)_{(-2)}\ \overline{G_{\beta a}}(z,\tfrac{1}{2}+ir,\chi)].$$

From the fact that $\Phi(s,\chi)_{(n)} = \dfrac{(-1)^{n/2}\,\Gamma(s)^2}{\Gamma(s+\frac{n}{2})\Gamma(s-\frac{n}{2})}\,\Phi(s,\chi)$, we see that the first term is $[\frac{1}{1/2+ir}]\ y + y^{1+2ir}\ \overline{\phi_{aa}}\ (\frac{1}{2}+ir,\chi) + y^{1-2ir}[\frac{1/2-ir}{1/2+ir}] \times \varphi_{aa}(\frac{1}{2}+ir,\chi)$.

The second term is small as $y \to \infty$, so we set:

(3.12) $I^+_{a,3}(s) =$
$$= \int_0^\infty y^{s-2}\ \mathrm{Im}\{\frac{1}{4\pi}\int_{-\infty}^{\infty} \theta_a^0[G_{\beta a}(\cdot,\tfrac{1}{2}+ir,\chi)_{(-2)}\ \overline{G_{\beta a}}(\cdot,\tfrac{1}{2}+ir,\chi)]\ S_2\varphi(\tfrac{1}{2}+ir)\ dr\}\ dy.$$

Next, we note that
$$\theta_a^0\ [E^*_\beta(\cdot,s_j,\chi)_{(-2)}E^*_\delta(\cdot,s_j,\chi)]$$
$$= [\varphi^*_{\beta a}(s_j,\chi)_{(-2)}y^{1-s_j}]\,[\overline{\varphi^*_{\delta a}}(s_j,\chi)y^{1-s_j}] + \theta_a^0[G^*_{\beta a}(\cdot,s_j,\chi)_{(-2)}G^*_{\delta a}(\cdot,s_j,\chi)]$$

Hence $\theta_a^0\ [K_{res}(\cdot,\cdot,\chi)](y) = \sum_{\text{poles } s_j} S_2\varphi(s_j)y^{2-2s_j}\ \varphi^*_{aa}(s_j,\chi)_{(-2)} + (\text{small})$.

We separate out the small part and define:

(3.13) $I^+_{a,4}(s)$
$$= \int_0^\infty y^{s-2}\ \mathrm{Im}\{\sum_{\text{poles } s_j} S_2\varphi(s_j) \sum_{\beta,\delta=1}^{h(\chi)} [\Phi^*(s_j,\bar\chi)^{-1}]_{\beta\delta}\ \theta_a^0[G^*_{\beta a,(-2)},\overline{G^*_{\delta a}})(y)\}dy.$$

This leaves us with the non-trivial term:

(3.14) $I^+_{\alpha,2}(s)$

$$= \int_0^\infty y^{s-2}\, \mathrm{Im}\{[\sum_n e^{2\pi i n a} k(z,z+n) - \frac{y}{4\pi}\int_{-\infty}^{\infty} \frac{S_2\varphi(\frac{1}{2}+ir)}{\frac{1}{2}+ir}\, dr]$$

$$- \frac{1}{4\pi}\int_{-\infty}^{\infty} [y^{1+2ir}\, \bar{\varphi}_{\alpha\alpha}(\tfrac{1}{2}+ir,\chi) + y^{1-2ir}[\tfrac{1/2-ir}{1/2+ir}]\, \varphi_{\alpha\alpha}(\tfrac{1}{2}+ir,\chi)\, S_2\varphi(\tfrac{1}{2}+ir)]\, dr$$

$$- \sum_{\text{poles } s_j} S_2\varphi(s_j) y^{2-2s_j}\, \varphi^*_{\alpha\alpha}(s_j,\chi)_{(-2)}\}\, dy$$

We next evaluate the $I^+_{\alpha,j}$.

(1) For $\beta \neq \alpha$,

$$I^+_{\alpha\beta,1}(s,\chi,\varphi) = \mathcal{I}m \sum_{n\neq 0} e^{2\pi i n a(\beta)} \sum_{\tau\in\Gamma_\beta\backslash\Gamma} (\int_{\Gamma\backslash G} E_\alpha(x,s,\chi)_{(2)}\, \varphi(x^{-1}\tau^{-1}\gamma^n_{\beta 0}\tau x)\, dx)$$

$$= \mathcal{I}m \sum_{n\neq 0} e^{2\pi i n a(\beta)} \int_{\Gamma_\beta\backslash G} E_\alpha(x,s,\chi)_{(2)}\, \varphi(x^{-1}\gamma^n_{\beta 0}x)\, dx$$

$$= \mathcal{I}m \sum_{n\neq 0} e^{2\pi i n a(\beta)} \int_{\Gamma_\infty\backslash G} E_\alpha(\sigma_\beta x,x,s)_{(2)}\, \varphi(x^{-1}\sigma_\beta^{-1}\gamma^n_{\beta 0}\sigma_\beta x)\, dx.$$

Now $\varphi \in S_{2,0}$ implies $\varphi(g) = e^{i\theta}\varphi(\mathrm{pol}\, g)$ where $g = y^{-1/2}k_\theta\begin{pmatrix} y & x\\ 0 & 1\end{pmatrix}$, and where $\mathrm{pol}(g) = (g^t g)^{1/2}$ is the polar part. If $g = \begin{bmatrix} g_{11} & g_{12}\\ g_{21} & g_{22}\end{bmatrix}$ then $e^{i\theta} = \frac{g_{11} - ig_{21}}{|g_{11} - ig_{21}|}$. Moreover if $p_z \in P = AN$ is the unique element so that $p_z i = z$, then $\varphi[\mathrm{pol}(p_z^{-1} g p_z)] = \phi[\frac{|z - z'|}{y^2}]$ where $gp_z i = z'$, and $\phi$ is a function on $\mathbb{R}^+$. We conclude: if $\alpha = \begin{pmatrix} a & b\\ c & d\end{pmatrix}$ then

$$\varphi(p_z^{-1}\alpha p_z) = \frac{(a - cx) - icy}{|(a - cx) - icy|}\, \phi[\frac{|z - \alpha z|^2}{y(z)\, y(\alpha z)}].$$

Now $\Gamma_\infty \subset P$ so $\int_{\Gamma_\infty\backslash G} E_\alpha(\sigma_\beta x,x,s)_{(2)}\, \varphi(x^{-1}\sigma_\beta^{-1}\gamma^n_{\beta 0}\sigma_\beta x)\, dx$

$$= \int_{\Gamma_\infty\backslash P} E_\alpha(\sigma_\beta p,s)_{(2)}\, \varphi[p^{-1}\begin{pmatrix} 1 & n\\ 0 & 1\end{pmatrix}p]\, dp$$

$$= \int_{\Gamma_\infty\backslash P} E_\alpha(\sigma_\beta p,s)_{(2)}\ \phi[\frac{n^2}{y^2}]\ dp$$

$$= \varphi_{\alpha\beta}(s,\chi)_{(2)} \int_0^\infty y^{1-s}\ \phi[\frac{n^2}{y^2}]\ \frac{dy}{y^2}$$

$$= \varphi_{\alpha\beta}(s,\chi)_{(2)}\ M\phi(s/2)\ |n|^{-s}.$$

Consequently,

$$I^+_{\alpha\beta,1}(s,\chi,\varphi) = [\sum_{n\neq 0} \frac{e^{2\pi i n a(\beta)}}{|n|^s}]\ M\varphi(s/2)\ \frac{1}{2i}\ [\varphi_{\alpha\beta}(s,\chi)_{(2)} - \varphi_{\alpha\beta}(s,\chi)_{(-2)}] = 0,$$

since $\varphi_{\alpha\beta}(\ )_{(n)} = \varphi_{\alpha\beta}(\ )_{(-n)}$. A similar argument holds if $\alpha = \beta$.

(2) As in (1), $k(z,z-n) = \phi[\frac{n^2}{y^2}]$ so we may extract from $I^+_{\alpha,2}$ the piece $\mathit{Im}$ $\int_0^\infty y^{s-2}\ [\sum_n e^{2\pi i n a}\ \phi[\frac{n^2}{y^2}] - \frac{y}{4\pi} \int_{-\infty}^{\infty} \frac{S_2\varphi(\frac{1}{2}+ir)}{\frac{1}{2}+ir}\ dr]$. Once again, because $\varphi(s,\chi)_{(2)}$ $= \varphi(s,\chi)_{(-2)}$, $S_2\varphi = S_{-2}\varphi^*$ and $\phi = \phi^*$, the integral equals its conjugate and makes no contribution to $I^+_{\alpha,2}$. We may rewrite the rest of $I^+_{\alpha,2}$ as $\mathit{Im}$ of:

$$-\int_0^\infty y^{s-2}\ \{\frac{1}{4\pi i} \int_{Re\tau=1/2} \varphi_{\alpha\alpha}(\tau,x)\ y^{2-2\tau}[\frac{1-\tau}{\tau}\ S_2\varphi(\tau) - S_2\varphi(1-\tau)]\ d\tau\}\ dy$$

Now $S_2\varphi(1/2-ir) = \frac{\frac{1}{2}+ir}{\frac{1}{2}-ir}\ S_2\varphi(1/2+ir)$ (cf. [He]), so we get:

$$-\int_0^\infty y^{s-2}\ \{\frac{1}{2\pi i} \int_{Re\tau=1/2} \varphi_{\alpha\alpha}(\tau,x)_{(-2)}\ y^{2-2\tau}\ S_2\varphi(\tau)\ dt\}\ dy$$

The contour shifting argument of (3A) applies here, and we get that

$I^+_{a,2}(s,\chi,\varphi) = \mathcal{I}m[\varphi_{aa}[\frac{s}{2}1,\ \chi]_{(-2)}\ S_2\varphi[\frac{s}{2}1]] = 0$ since again $\varphi_{aa}(\cdot)_{-2} = \varphi_{aa}(\cdot)_2$ and $S_2\varphi = S_2^*\varphi^*$.

We conclude that $I^+_{a,2} \equiv 0$.

(3) Next consider $I^+_{a\beta,3}(s)$. Recall from the Appendix that

$$E_\beta(x,s)_{(n)} = a_{0\beta a}(y,s)_n + G_{\beta a}(x,s)_n,$$
$$G_{\beta a}(x,s)_n = y^{1-s} \sum_{m\neq 0} \varphi_{\beta am}(s)\ W_{s,n}(2\pi|m|y)$$

with $W_{s,n}(y) = \int_{-\infty}^{\infty} [\frac{t+i}{|t+i|}]^n(1+t^2)^{-s}\ e(-ty)\ dt$, (similarly for $E_\beta(x,s,\chi)$).

So $\int_0^\infty y^{s-2}\theta_a^0\ [G_{\beta a}(\cdot,\frac{1}{2}+ir,\chi)_{(-2)}\ \overline{G_{\beta a}}(\cdot,\frac{1}{2}+ir,\chi)]\ dy$

$$= [\sum_{m\neq 0} \frac{|\phi_{\beta am}(\frac{1}{2}+ir,\chi)|^2}{|m|^s}] \int_0^\infty y^s W_{1/2+ir,-2}(2\pi y)\ W_{1/2+ir,0}(2\pi y)\ \frac{dy}{y}$$
$$= R_{a\beta}(\frac{1}{2}+ir,s,\chi)\ (2\pi)^{-s} \int_0^\infty y^s W_{1/2+ir,-2}(y)\ W_{1/2+ir,0}(y)\ \frac{dy}{y}.$$

Let $A^0_{-2}(\frac{1}{2}+ir,s)$ denote the integral. Taking the conjugate amounts to replacing $W_{s,-2}$ by $W_{s,2}$.

So $I^+_{a\beta,3}(s) = [\frac{1}{4\pi} \int_{-\infty}^{\infty} R_{a\beta}((\frac{1}{2}+ir,s,\chi)A^+(\frac{1}{2}+ir,s)\ S\varphi(\frac{1}{2}\ ir)\ dr]$,

where $A^+ = \frac{1}{2i}(A_{-2} - A_2)$. It is not hard to evaluate $A^+(\tau)$, and to see that the integral can be analytically continued. First,

$$\frac{1}{2i}[W_{\tau,n}(y) - W_{\tau,-n}(y)] = \int_{-\infty}^{\infty} \mathrm{Im}[\frac{t+i}{|t+i|}]^n(1+t^2)^{-\tau}\ e(-ty)\ dt.$$

For n = 2, $\mathcal{I}m W_{\tau,n}(y) = -2\frac{\partial}{\partial y}\int_{-\infty}^{\infty}(1+t^2)^{-\tau}\, e(-ty)\, dt = -2\frac{\partial}{\partial y} W_{\tau+1,0}(y)$.

So $A^+(\frac{1}{2}+ir,s) = -2\int_0^{\infty} y^s[\frac{\partial}{\partial y} W_{3/2+ir}(y)]\,[\overline{W_{1/2+ir}(y)}]\,\frac{dy}{y}$. Recall that

$$W_\tau(|y|) = (2\pi)^\tau |y|^{\tau-1/2}\Gamma(\tau)\, K_{\tau-1/2}(2\pi|y|),$$

that $\frac{\partial}{\partial y} K_\nu(y) = -\frac{1}{2}[K_{\nu-1}(y) + K_{\nu+1}(y)]$ (G-R, p. 970),

and that $\int_0^{\infty} x^{-\lambda} K_\mu(2\pi x)\, K_\nu(2\pi x)\, dx =$

$$= \frac{2^{-2-\lambda}}{\Gamma(1-\lambda)}\,\Gamma[\tfrac{1-\lambda+\mu+\nu}{2}]\,\Gamma[\tfrac{1-\lambda-\mu+\nu}{2}]\,\Gamma[\tfrac{1-\lambda+\mu-\nu}{2}]\,\Gamma[\tfrac{1-\lambda-\mu-\nu}{2}] \qquad \text{(G-R, p. 693)}.$$

The result is:

$$\begin{aligned} A^+(\tfrac{1}{2}+ir,s) = {} & (-2)(2\pi)^2\,\Gamma(3/2+ir)\,\overline{\Gamma(1/2+ir)}\,(2\pi)^{-s+2}\,\frac{2^{-2+s-1}}{\Gamma(s)} \\ & \cdot \{(1+ir)\,\Gamma(s+1)\,\Gamma(s-1)\,\Gamma(s-1-2ir) \\ & \times \Gamma(s+1+2ir) - 1/2\,[\Gamma(s)^2\,\Gamma(s-2ir)\,\Gamma(s+2ir) \\ & + \Gamma(s+2)\,\Gamma(s-2-2ir)\,\Gamma(s+2+2ir)\,\Gamma(s-2)\}. \end{aligned}$$

Every feature of the analytic continuation of $I^+_{\alpha\beta}(s)$ can therefore be reduced to the case (3A).

(C) The last trace formula we need to spell out are those for $\mathrm{Tr}\ \pi_0\sigma R^\Gamma_\varphi\pi_0$, where $\sigma$ is a residue of $E_s$ or $X_+E_s$. Of course these follow immediately by taking the residues of the terms in the trace formulae for $\mathrm{Tr}\ \pi_0 E_s R^\Gamma_\varphi \pi_0$ (resp. $X_+E_s$); this indeed is a principal advantage of having an analytic continuation for all terms. The details may be left to the reader.

## § 4. Spectral Estimates

In this section we are concerned with the spectral functions $N(\sigma,\chi,T)$ and $M(\sigma,\chi,T)$. We recall (from the Appendix) that these are defined by:

(4.1) Definition. For cuspidal $\sigma$,

(i) $N(\sigma,\chi,T) = \sum_{0\le|t_j|\le T} \langle Op(\sigma)u_j^\chi,\ u_j^\chi\rangle$

(ii) $M(\sigma,\chi,T) = -\frac{1}{4\pi}\int_{-T}^{T} \langle Op(\sigma)E(\cdot,\tfrac{1}{2}+ir,\chi),\ E(\cdot,\tfrac{1}{2}+ir,\chi)\rangle\, dr$

For Eisenstein $\sigma = E_{a,s}$ replace (ii) by:

(ii)' $M(E_{a,s},\chi,T) = -\frac{1}{4\pi}\sum_{\beta=1}^{h(\chi)}\int_{-T}^{T} R_a[|E_\beta(\cdot,\tfrac{1}{2}+ir,\chi)|^2,\ s]\, dr;$

for $\sigma = X_+E_{a,s}$, replace $R_a$ by $R_a^+$, where

(ii)" $R_a^+(|E_\beta(\cdot,\tfrac{1}{2}+ir,\chi)|^2,\ s) = A^+(\tfrac{1}{2}+ir,s)\ R_{a\beta}(\tfrac{1}{2}+ir,s,\chi)$

(cf. (§3B.3)).

For residual $\sigma$, one takes the residue of the corresponding $M$ or $N$ function.

$M(E_{a,s},\chi,T)$ is the finite part or renormalization (in the sense of [Za1]) of the divergent expression

$$\sum_{\beta=1}^{h(\chi)}\int_{-T}^{T} \langle E_a(\cdot,s)\ E_\beta(\cdot,\tfrac{1}{2}+ir,\chi),\ E_\beta(\cdot,\tfrac{1}{2}+ir,\chi)\rangle\, dr.$$

When $\sigma \equiv 1$, $\chi \equiv 1$, the spectral functions are;

(4.2) (i) $N_\Gamma(T) = \sum_{|t_j|\le T} 1$

$$\text{(ii)} \quad M_\Gamma(T) = -\frac{1}{4\pi}\int_{-T}^{T} \frac{\Delta'}{\Delta}\left(\tfrac{1}{2}+ir\right) dr$$

where $\Delta(s) = \det \Phi(s)$. It is a classical result of the trace formula that:

$$(4.3) \quad (M_\Gamma + N_\Gamma)(T) = \frac{|\Gamma|}{4\pi} T^2 - \frac{h}{4\pi} \ln T + C_\Gamma T + O(T/\ln T) \qquad \text{(cf. [V])}.$$

No separate estimates exist in general on $N_\Gamma$ or $M_\Gamma$; indeed they would presumably depend critically on $\Gamma$ (cf. [PS]). Despite this dependence, we henceforth drop the subscript $\Gamma$ from the notation, for simplicity.

Our main purpose in this section is to provide analogues of the Weyl law (4.3) for the spectral functions $N(\sigma,\chi,T) + M(\sigma,\chi,T)$. We also give crude estimates in the corresponding alsolute sums $(|N| + |M|)((\sigma,\chi,T)$.

The generalized Weyl laws are:

<u>Theorem 4A</u>. For cuspidal $\sigma$, $(N+M)(\sigma,\chi,T) \underset{(s,m)}{<<} T/\ln T$.

<u>Theorem 4B</u>. For Eisenstein $\sigma$, on $\mathrm{Re}\, s = \frac{1}{2}$, and for residual $\sigma$, $(N+M)(\sigma,\chi,T) \underset{(s,m)}{<<} T^{3/2}$.

The order relation may depend on the $(s,m)$ parameters on $\sigma$. A tedious but straightforward argument shows that the dependence is polynomial.

Theorem 4A is the direct analogue of the Weyl law for the compact case (1.7). Theorem 4B has no compact analogue. It partly corroborates the mean Lindelof hypothesis for $\Gamma = SL_2(\mathbb{Z})$:

$$(4.4) \qquad \text{(MLH)} \quad \sum_{|t_j|\le T} |\langle E_s u_j, u_j\rangle| << |s|^A\, T^{3/2+\epsilon} \quad (\mathrm{Re}\, s = \tfrac{1}{2}),$$

except that it is only an estimate on the non-absolute sum.

For the absolute sums $|M|$ and $|N|$ (put absolute values around the matrix elements in (4.1)), we are going to need the following estimates:

Proposition 4C. For all $\sigma$ above,

$$(|M| + |N|)(\sigma,\chi,T) << (M_\Gamma + N_\Gamma)(T).$$

The proofs of Theorems A and B follow very closely the pattern of the compact case in [Z3]. In fact, we will give full details only on the new aspects of the estimates. As explained in § 1, detailed comparison with [Z3] will require some transcribing of the trace formulae of § 2 - § 3 into the notation of [Z3], i.e. in terms of the operators $Op(\chi\sigma)\hat{f}(R)$.

We begin with the proof of Theorem 4A.

Proof. (a) Suppose first $\sigma$ has weight 0, so $\sigma = u_k$. The trace formula then reads:

$$(4.5) \quad \sum_{n=1}^{\infty} \langle u_k u_n^\chi, u_n^\chi \rangle \hat{f}(r_n) + \frac{1}{4\pi} \sum_{\beta=1}^{h(\chi)} \int_{-\infty}^{\infty} \langle u_k E_\beta(\cdot,\tfrac{1}{2}+ir,\chi), E_\beta(\cdot,\tfrac{1}{2}+ir,\chi)\rangle \hat{f}(r)\, dr$$
$$= \sum_{\{\gamma\}_{hyp}} \chi(\gamma) \left(\int_{\gamma_0} u_k\right) M_k f(L_\gamma) + (ell) + (res),$$

where

$$(ell) = \sum_{\{\gamma\}_{ell}} \chi(\gamma)\, (\nu(\gamma_0))^{-1} u_k(z_\gamma)\, L_k f(k_\gamma),$$

where (res) is the finite trace on the residual subspaces, and, most importantly, where

$$(4.6) \qquad M_k\, f(\ell) = f(\ell) + \int_\ell^\infty f(\ell)\, \frac{\partial}{\partial t}\, F_k\, \{[-(\frac{\mathrm{ch}t - \mathrm{ch}\ell}{(\mathrm{sh}\frac{\ell}{2})^2}]\}\, dt$$

wuth $F_k(x) = F(\frac{1}{4} + \frac{s_k}{4}, \frac{1}{4} - \frac{s_k}{4}, 1, x)$ (see [Z3], Theorem 4.1).

As in ([Z3], Theorem 5.1), we let $\hat{f}_T(r) = \chi_T((1 + r^2)^{1/2})$, $\chi_T$ being the characteristic function of $[0,T]$; and $\hat{f}_{T,\epsilon} = \hat{f}_T * \psi_\epsilon$, where $\psi_\epsilon$ is an approximate identity satisfying $\int\psi = 1$, $\psi$ even and supp $\psi \subset [-1, 1]$. (Note: $\hat{f}$ is written f in [Z3].)

Exactly as in the case $\sigma \equiv 1$, the residual and elliptic terms are bounded as $\epsilon \downarrow 0$, $T \longrightarrow \infty$ (since $u_k$ is bounded).

Since $f_{T,\epsilon} \in C_0^\infty$, (4.5) applies and we have:

$$(4.7) \qquad (M + N)\,(u_k,\chi,T) = \sum_{n=1}^{\infty} \langle u_k u_n^\chi, u_n^\chi\rangle\, (\hat{f}_T - \hat{f}_{T,\epsilon})(r_n)$$
$$+ \sum_{\beta=1}^{h(\chi)} \frac{1}{4\pi} \int_{-\infty}^{\infty} \langle u_k E_\beta(\cdot,\tfrac{1}{2}+ir,\chi)\, E_\beta(\cdot,\tfrac{1}{2}+ir,\chi)\rangle\, (\hat{f}_T - \hat{f}_{T,\epsilon})(r)\, dr$$
$$+ \sum_{\{\gamma\}_{hyp}} \chi(\gamma)\, (\int_{\gamma_0} u_k)\, M_k f_{T,\epsilon}(L_\gamma) + O(1).$$

But the hyperbolic term is precisely what appears in the compact case ([Z3], p. 325). Letting $\epsilon = \frac{a}{\ln T}$, for $a >> 1$, one finds that the hyperbolic term is $O(T^\delta)$ for any $\delta > 0$ (in [Z3] we set $a = 1$ and $\delta = \frac{1}{2}$).

This leaves the spectral term. What is needed to conclude the proof, exactly as in [Z3], is the estimate.

$$(4.8)\quad \sum_{T\le t_j\le T+\epsilon} |\langle u_k u_n^\chi, u_n^\chi\rangle| + \frac{1}{4\pi}\sum_{\beta=1}^{h(\chi)} \int_T^{T+\epsilon} dr\, |\langle u_k E_\beta(\cdot,\tfrac{1}{2}+ir,\chi), E_\beta(\cdot,\tfrac{1}{2}+ir,\chi)\rangle|$$
$$<< \epsilon T + O(T/\ln T).$$

This can be reduced to the standard estimate:

$$(4.9)\qquad (N_\Gamma + M_\Gamma)(T+\epsilon) - (N_\Gamma + M_\Gamma)(T) << \epsilon T + O(T/\ln T),$$

which follows from the Weyl law (4.3). Indeed, we use the simple bound $|\langle u_k u_n^\chi, u_n^\chi\rangle| << 1$ in the first term of (4.8). For the second we use the bound:

$$(4.10)\quad \sum_{\beta=1}^{h(\chi)} |\langle u_k E_\beta(\cdot,\tfrac{1}{2}+ir,\chi), E_\beta(\cdot,\tfrac{1}{2}+ir,\chi)\rangle| << |\frac{\Delta'}{\Delta}(\tfrac{1}{2}+ir)| = -\frac{\Delta'}{\Delta}(\tfrac{1}{2}+ir) + O(1).$$

The estimate (4.10) comes from the Maass-Selberg relation (cf. [C-S], 7.12; [V], 4.3.1). We sketch the proof, pretending that $\chi \equiv 1$ and $h = 1$, with the cusp at $\infty$.

First, we write:

$$(4.11)\qquad \langle u_k E(\cdot,\tfrac{1}{2}+ir), E(\cdot,\tfrac{1}{2}+ir)\rangle$$
$$= \langle u_k \tilde{E}_A(\cdot,\tfrac{1}{2}+ir), \tilde{E}_A(\cdot,\tfrac{1}{2}+ir)\rangle + \int_{y\ge A} u_k |a_0(iy, \tfrac{1}{2}\, ir)|^2\, dz$$
$$+ \int_{y\ge A} u_k (2\,\mathrm{Re}\, a_0\cdot G)\, dz.$$

The first term is $O(\|u_k\|_\infty \|\tilde{E}_A(\cdot,\tfrac{1}{2}+ir)\|^2)$. The second is bounded by a constant independent of r. The third is $O_k(\int_{y\ge A} |G(\cdot,\tfrac{1}{2}+ir)|^2\, dz)^{1/2}$ by Shwarz, hence is $O(\|\tilde{E}_A(\cdot,\tfrac{1}{2}+ir)\|)$. It folows that:

$$(4.12)\qquad |\langle u_k E(\cdot,\tfrac{1}{2}+ir),\ E(\cdot,\tfrac{1}{2}+ir)\rangle| = O(\|\tilde{E}_A(\cdot,\tfrac{1}{2}+ir)\|)$$
$$= O(|\tfrac{\Delta'}{\Delta}(\tfrac{1}{2}+ir)|)$$
$$= O(-\tfrac{\Delta'}{\Delta}(\tfrac{1}{2}+ir)) + O(1)$$

(cf. [C-S], 7.12).

It follows that the left side of (4.8) is bounded by the left side of (4.9), concluding the proof.

(b) Now suppose $\sigma = X_+u_k$ or $\sigma = \psi_m$.

The trace formulae now have the form:

$$(4.13)\qquad \sum_n \langle Op(\sigma)u_n^\chi,\ u_n^\chi,\rangle\ J_m\chi_1(r_n^\chi)\hat{f}(r_n^\chi)$$
$$+ \frac{1}{4\pi}\sum_{\beta=1}^{h(\chi)} \int_{-\infty}^{\infty} \langle Op(\sigma)E_\beta(\cdot,\tfrac{1}{2}+ir,\chi)\ E_\beta(\cdot,\tfrac{1}{2}+ir,\chi)\rangle\ J_m\chi_1(r)\hat{f}(r)\,dr$$
$$= (hyp) + (ell) + (res),$$

where $m = 2$ if $\sigma = X_+u_k$; $J_m\chi_1$ is a cut-off (no relation to the character $\chi$) as in [Z3], § 1 (f); where (ell) + (res) are finite sums as in § 2; and where the significant term (hyp) is given by:

$$(4.14)\qquad (hyp) = \sum_{\{\gamma\}_{hyp}} \chi(\gamma)(\textstyle\int_{\gamma_0}\sigma)\ P_m^{\chi_1}\ f(L_\gamma)$$

where $P_m^{\chi_1}$ is the transform in ([Z3], p. 328 and p. 332).

We now use the same test function $\hat{f}_{T,\epsilon}$ as in (a). All terms therefore depend on $(T,\epsilon)$. Setting $\epsilon = a/\ln T$ as before, $(hyp)_{T,\epsilon} = O(T^\delta)$ by [Z3], § 5. The proof is concluded with an estimate like (4.8) except with $Op(\sigma)$ in place of $u_k$. The bound $\|Op(\sigma)\| << 1$ applies again to the discrete sum. So the main point is the analogue of (4.10). This goes as in the weight 0 case, once we recall how $Op(\sigma)$ operates on $E_\beta(\cdot,\frac{1}{2}+ir,\chi)$ ([Z3], § 1e). Identifying G with $D \times B$ ($B = \text{bd } D$) as in ([Z3], § 1c), $\sigma(z,1)$ become a classical automorphic form of weight m on $D(\sigma(\gamma z,1)) = j(\gamma,z)^{-2m}\,\sigma(z,1)$. Then:

$$(4.15) \qquad Op(\sigma)\, E_\beta(\cdot,\tfrac{1}{2}+ir,\chi) = \sigma(z,1)\, L_m^-\, E_\beta(\cdot,\tfrac{1}{2}+ir,\chi)$$

where $L_m^-$ is the unitary lowering operator by m units ([Z3], § 1e). Thus, $L_m^-\, E_\beta(\cdot,\chi,s) = E_\beta(\cdot,s,\chi)_{(-m)}$, giving (4.15).

Thus, the estimate we need is:

$$(4.16) \qquad \langle \sigma(\cdot,1)E_\beta(\cdot,\tfrac{1}{2}+ir)_{(-m)},\, E_\beta(\cdot,\tfrac{1}{2}+ir)\rangle << \left|\frac{\Delta'}{\Delta}(\tfrac{1}{2}+ir)\right|$$

(again pretending $\chi \equiv 1$, $h = 1$ and with the cusp at $\infty$).

Exactly as in the weight 0 case, the main point of the estimate is:

$$(4.17) \qquad \langle \sigma(\cdot,1)\tilde{E}_A(\cdot,\tfrac{1}{2}+ir)_{(-m)},\, \tilde{E}_A(\cdot,\tfrac{1}{2}+ir)\rangle << \left|\frac{\Delta'}{\Delta}(\tfrac{1}{2}+ir)\right|$$

By Schwarz, this follows as long as:

$$(4.18) \qquad \|\tilde{E}_A(\cdot,\tfrac{1}{2}+ir)_{(-m)}\|^2 + \|\tilde{E}_A(\cdot,\tfrac{1}{2}+ir)\|^2 << \left|\frac{\Delta'}{\Delta}(\tfrac{1}{2}+ir)\right|$$

The Maass-Selberg estimate for weight 0 has a straightforward generalization to weight m:

$$(4.19)\qquad \|\tilde{E}_A(\cdot,\tfrac{1}{2}+ir)_{(-m)}\|^2 = -\frac{\Delta_m'}{\Delta_m}(\tfrac{1}{2}+ir) + O(1),$$

where $\Delta_m(s)$ is the scattering determinant for weight m Eisenstein series. One has:

$$(4.20)\qquad \Delta_m(s) = C_m\left[\frac{\Gamma(s)\,\Gamma(s-\frac{1}{2})}{\Gamma(s+\frac{m}{2})\,\Gamma(s-\frac{m}{2})}\right]^{h(\chi)} \Delta(s)$$

(see the Appendix). So one gets new terms of the form $\frac{\Gamma'(z)}{\Gamma(z)}$ with Re $z = 0, \frac{1}{2}, \frac{m+1}{2}, -(\frac{m+1}{2})$. But $\frac{\Gamma'}{\Gamma}(z) = \ln z - \frac{1}{2z} + O(|z|^{-2})$ on $|\arg z| \le \pi - \delta$, $\delta > 0$, $|z| > \delta$ ([Iv], p. 492). So the new terms are negligible. The proof is concluded as before. ■

We now turn to the proof of Theorem B.

Proof. The main new complicaiton is that Eisenstein series are unbounded as $|y| \to \infty$, so the Maass-Selberg estimates (4.9) and (4.15) and the simple bounds $\|Op(\sigma)\| \ll 1$ no longer apply. The hyperbolic term also changes, but as we shall see, only in a minor way. Furthermore, we have some new terms in the trace formula.

First, assume as in (a) that $\sigma$ has weight 0, i.e. $\sigma = E_{\alpha,s}$. To replace $|\langle u_k u_n^\chi, u_n^\chi\rangle| \ll 1$, we claim:

$$(4.21)\qquad |\langle E_{\alpha,s}u_j, u_j\rangle| \ll_s |r_j|^{1/2} \ln|r_j| \quad (\mathrm{Re}\ s = 1/2).$$

Proof. $|\langle E_{\alpha,s}u_j,u_j\rangle| \le \int_{F_A} |E_\alpha(z,s)||u_j(z)|^2\, dz + \sum_\beta \int_{y_\beta \ge A}$ (same). Since

$E_\alpha(z,s) = \delta_{\alpha\beta} y_\beta^s + \phi_{\alpha\beta}(s)\, y_\beta^{1-s} + G_{\alpha\beta}(z,s)$ with $G_{\alpha\beta}(z,s)$ bounded in the cuspidal end $\{y_\beta \geq A\}$, it is clear that

(i) $|E_\alpha(z,s)| \ll_s A^{1/2}$ in $F_A$ (Re $s = 1/2$)

(ii) $\int_{y_\beta \geq A} |E_\alpha(z,s)|\, |u_j(z)|^2\, dz \ll_s A^{1/2}$

(iii) $\int_{y_\beta \geq A} |E_\alpha(z,s)|\, |u_j(z)|^2\, dz \ll_s \int_{y_\beta \geq A} y_\beta^{1/2}\, |u_j(z)|^2\, dz$

Further,

$$\int_{y_\beta \geq A} y_\beta^{1/2}\, |u_j(z)|^2\, dz = \sum_{n \neq 0} |\rho_{j\beta}(n)|^2 \int_A^\infty y^{1/2} K_{ir_j}(2\pi|n|y)^2\, dy.$$

The following estimate on $|\rho_{j\beta}(n)|^2$ appears in [D-I], Lemma 4:

$$\sum_\beta |\rho_{j\beta}(n)|^2 \ll (|n| + |r_j|) \ln (2 + |n|\, |r_j|^{-1}) \operatorname{ch}\pi|r_j|^{-1}$$

This estimate, stated in [D-I] for $\Gamma = PSL_2(\mathbb{Z})$ holds for general cofinite $\Gamma$ (with the same proof).

Continuing,

$$\sum_\beta \sum_{n \neq 0} |\rho_{j\beta}(n)|^2 \int_A^\infty y^{-1/2} K_{ir_j}(2\pi|n|y)^2\, dy.$$

$$\ll A^{-1/2} \operatorname{ch}\pi r_j \sum_{n \neq 0} (|n| + |r_j|) \ln (2 + |n|\, |r_j|^{-1}) \int_A^\infty K_{ir_j}(2\pi|n|y)^2\, dy$$

Following ([I], Lemma 7), we break up $\sum_{n \neq 0}$ into $\sum_{|n| \leq 2|r_j|}$ and $\sum_{|n| \geq 2|r_j|}$. For the first term, $\int_A^\infty K_{ir_j}(2\pi|n|y)^2\, dy$ is estimated by $\int_0^\infty(\text{same}) = n^{-1}(\operatorname{ch}\pi|r_j|)^{-1}$.

The first term accordingly contributes no more than

$$A^{-1/2} \sum_{|n|\leq 2|r_j|} (1 + |r_j|/|n|) \ll A^{-1/2} |r_j| (1 + \ln|r_j|).$$

For the second, one uses $|K_{ir_j}(y)| \ll e^{(\pi/2)|r_j|-y}$ so that $\int_A^\infty K_{ir_j}(2\pi|n|y)^2 \, dy \ll e^{\pi|r_j|-4} e^{-4\pi|n|A}$. The second term contributes

$$A^{-1/2} e^{2\pi|r_j|} \sum_{|n|\leq 2|r_j|} \ln (2 + |n||r_j|^{-1}) e^{-4\pi|n|A} \ll A^{-1/2} e^{2\pi|r_j|(1-4\pi A)}$$

for $A \gg 1$.

All told,

$$|\langle E_{a,s}u_j,u_j\rangle| \lesssim_s A^{1/2} + A^{1/2} |r_j|(1 + \ln|r_j|) + A^{-1/2} e^{2\pi|r_j|(1-4\pi A)} + O(1).$$

Setting $A = |r_j|$ gives the lemma.

(4.22) <u>Corollary</u>. $\sum_{T\leq|r_j|\leq T+\epsilon} |\langle E_{a,s}u_j,u_j\rangle| \ll T^{3/2}$ if $\epsilon = a/\ln T$ and $\mathrm{Re}\, s = 1/2$.

We need a similar bound on $\int_T^{T+\epsilon} |R_a[|E_\beta|^2,s]| \, dr$ on $\mathrm{Re}\, s = 1/2$, $\epsilon = \cdots$

(4.23) <u>Claim</u>. $\sum_{\beta=1}^{h} R_a[|E_\beta(\cdot,\tfrac{1}{2}+ir)|^2,s] \lesssim_s |r|^{1/2} (\ln|r|) [\frac{\Delta'}{\Delta}(\tfrac{1}{2}+ir)]$

Proof. By (3.85), and setting $s = \frac{1}{2}+it$,

$$R_\alpha[|E_\beta|^2,s] = \int_{F_A} E_\alpha(z,\tfrac{1}{2}+it)\ |E_\alpha(z,\tfrac{1}{2}+ir)|^2\ dz - h_{\beta\alpha A}(\tfrac{1}{2}+it)$$
$$- \sum_\delta \phi_{\alpha\delta}(\tfrac{1}{2}+ir)h_{\beta\delta A}(\tfrac{1}{2}+it)$$
$$+ \sum_\delta \int_{y_\delta \geq A} [E_\alpha(z,\tfrac{1}{2}+it)|E_\beta(z,\tfrac{1}{2}+ir)|^2 - A_{\alpha\delta}(y_\delta,\tfrac{1}{2}+it)\ |A_{\beta\delta}(y_\delta,\tfrac{1}{2}+ir)|^2]\ dz$$

$(h_{\beta\alpha A}(s) = \int_0^A y^{s-2}\ |A_{\beta\alpha}|^2\ dy)$.

It is immediate that the $h_{\beta\alpha A}$ terms are $O(A^{1/2})$.

The $\int_{F_A}$ we bound as before by

$$[O(1) + A^{1/2}] \int_{F_A} |E_\beta(z,\tfrac{1}{2}+ir)|^2\ dz \ll (1 + A^{1/2}) \int_F |\tilde{E}_{\beta,A}(z,\tfrac{1}{2}+ir)|^2\ dz$$
$$= (1 + A^{1/2})\ [\ln A - \frac{\Delta'}{\Delta}(\tfrac{1}{2}+ir) + \mathrm{Re}\ \frac{\Delta(\frac{1}{2} - ir)}{r}\ \sin 2r \ln A]$$

(by the Maass-Selberg relations)

$$\ll (1 + A^{1/2})\ [\ln A - \frac{\Delta'}{\Delta}(\tfrac{1}{2}+ir)].$$

The term $\sum_\delta$ equals

$$\sum_\delta \int_{y_\delta \geq A} [A_{\alpha\delta}(z,\tfrac{1}{2}+it)|G_{\beta\delta}(z,\tfrac{1}{2}+ir)|^2$$
$$+ G_{\alpha\delta}(z,\tfrac{1}{2}+it)\ 2\ \mathrm{Re}\ A_{\beta\delta}\ \overline{G_{\beta\delta}}(z,\tfrac{1}{2}+ir) + G_{\alpha\delta}(z,\tfrac{1}{2}+it)\ |G_{\beta\delta}|^2]\ dz$$
$$\ll \sum_\delta \int_{y_\delta \geq A} (y_\delta^{1/2} + 1)\ |G_{\beta\delta}|^2\ dz$$
$$+ \sum_\delta [\int_{y_\delta \geq A} |G_{\alpha\delta}(z,\tfrac{1}{2}+it)|^2\ dz]^{1/2}\ (\int_{y_\delta \geq A} |A_{\beta\delta}\ \overline{G_{\beta\delta}}|^2\ dz)^{1/2}$$

$$<< \sum_\delta \int_{y_\delta \geq A} y_\delta^{1/2} |G_{\beta\delta}|^2 \, dz + (\int_{y_\delta \geq A} y_\delta |G_{\beta\delta}|^2 \, dz)^{1/2}$$

To estimate $\int_{y_\delta \geq A} y_\delta^{1/2} |G_{\beta\delta}|^2 \, dz$ and $\int_{y_\delta \geq A} y_\delta |G_{\beta\delta}|^2 \, dz$, we need the analogue for Fourier coefficients of Eisenstein series of the previous bound on $\rho_j(n)$. The bounds we need are on $\phi_{\beta\delta n}(\frac{1}{2}+ir)$ for $|n| \geq 1$. Following [D-I] again, we show that $|\phi_{\beta\delta n}(\frac{1}{2}+ir)| << (n+|r|) \ln (2 + n|r|^{-1}) \frac{\Delta'}{\Delta}(\frac{1}{2}+ir)$. Indeed $\phi_{\beta\delta n}(\frac{1}{2}+ir) \, n^{-1/2} \, W_{ir}(ny) = \int_0^1 \tilde{E}_{\beta,Y_0}(x + iy, \frac{1}{2}+ir) \, e(-nx) \, dx$ for $y \geq Y_0$ ($\kappa_\beta$ is assumed at $\infty$). One concludes:

$$|\phi_{\beta\delta n}(\tfrac{1}{2}+ir)|^2 << (n+|r|) \ln (2 + |r|^{-1}n) \, [\int_F |\tilde{E}_{\beta,Y_0}(z,\tfrac{1}{2}+ir)|^2 \, dz]^{1/2}$$
$$<< (n+|r|) \ln (2 + |r|^{-1}n) \, [\tfrac{\Delta'}{\Delta}(\tfrac{1}{2}+ir) + O(1)]$$

as $Y_0$ may be fixed independently of r.

Consequently,

$$\int_{y_\delta \geq A} y_\delta^{1/2} |G_{\beta\delta}|^2 \, dz = \operatorname{ch}\pi r \sum_{n\neq 0} |\phi_{\beta\delta n}(\tfrac{1}{2}+ir)|^2 \int_A^\infty y^{1/2} \, K_{ir}(2\pi|n|y)^2 \, dy$$
$$<< |r| \ln |r| \, \tfrac{\Delta'}{\Delta}(\tfrac{1}{2}+ir)$$

by the same argument as before. A small modification yields the same (or smaller) bound on the second term.

(4.24) Corollary. $\int_T^{T+\epsilon} |R_\alpha[|E_\beta(\cdot,\frac{1}{2}+ir,s)|^2, \frac{1}{2}+it]| \, dr << T^{3/2}$ if $\epsilon = \text{const.}/\ln T$.

Proof. It follows from the trace formula that $\int_T^{T+\epsilon} \frac{\Delta'}{\Delta}(\frac{1}{2}+ir)\, dr \ll T/\ln T$ if $\epsilon = \text{const.}/\ln T$.

We now complete the proof of the estimate:

Theorem 4B. $\sum_{|r_j|\geq T} \langle E_{\alpha,s}u_j,u_j\rangle - \sum_{\beta=1}^{h} \frac{1}{4\pi}\int_{-T}^{T} R_\alpha[|E_\beta(\cdot,\frac{1}{2}+ir)|^2,s]\, dr \ll T^{3/2}$ (Re $s = 1/2$).

Proof. This is now very similar to Theorem 4A(a), so we merely indicate the necessary modifications.

First,

$$\begin{aligned}(N + M)(E_{\alpha,s},\chi,T) &= \sum_{r_j} \langle E_{\alpha,s}u_j,u_j\rangle\, [\hat{f}_T - \hat{f}_{T,\epsilon}(r_j)] \\ &+ \sum_{\beta} (-\frac{1}{4\pi}) \int_{-\infty}^{\infty} R_\alpha[|E_\beta|^2,s](\hat{f}_T - \hat{f}_{T,\epsilon})(r)\, dr \\ &+ \sum_{\{\gamma\}_{hyp}} \chi(\gamma)\, (\int_{\gamma_0} E_{\alpha,s})\, M_s\, f(L_\gamma) + (\text{ell})_{T,\epsilon} \\ &+ \sum_{i\neq 3} I_{\alpha,i}(s,\chi,f_{T,\epsilon}).\end{aligned}$$

We handle the spectral sums as before. Thus,

$$\begin{aligned}&\sum_{r_j} |\langle E_{\alpha,s}u_j,u_j\rangle|\, |(\hat{f}_T - \hat{f}_{T,\epsilon}(r_j)| \\ &\ll \sum_{r_j\leq T-\epsilon} |\langle E_{\alpha,s}u_j,u_j\rangle|\, (\frac{T-r_j}{\epsilon})^{-N} + \sum_{T-\epsilon\leq r_j\leq T+\epsilon} |\langle E_{\alpha,s}u_j,u_j\rangle| \\ &\quad + \sum_{r_j\leq T-\epsilon} |\langle E_{\alpha,s}u_j,u_j\rangle|\, (\frac{T-r_j}{\epsilon})^{-N}\end{aligned}$$

$$= I + II + III.$$

The middle term is (for $\epsilon$ = const./ln T) $T^{3/2}$ (as noted above). The sum $I = \sum_{|r_j| \leq T-\epsilon}$ one splits up into the pieces $|r_j| < \epsilon$, $\epsilon < |r_j| < 2\epsilon$, $\epsilon k < r_j < (k+1)\epsilon, \cdots$

Using the bound

$$\left| \sum_{(k-1)\epsilon \leq r_j \leq k\epsilon} \langle E_{\alpha,s} u_j, u_j \rangle \left(\frac{T-r_j}{\epsilon}\right)^{-N} \right| << k^{-N}(k\epsilon)^{3/2}$$

one has, summing over k, $I << \epsilon^{3/2}$. The same argument shows that $II << \epsilon^{3/2}$. So

$$I + II + III << T^{3/2}.$$

Essentially the same argument shows that

$$\left| -\frac{1}{4\pi} \sum_{\beta} \int_{-\infty}^{\infty} |R_{\alpha}(|E_{\beta}|^2,s)|^2 \, (\hat{f}_T - \hat{f}_{T,\epsilon})(r) \, dr \right| << T^{3/2}.$$

One now breaks up the integral $\int_0^{\infty}$ into

$$\int_0^{T-\epsilon} + \int_{T-\epsilon}^{T+\epsilon} + \int_{T+\epsilon}^{\infty} \quad \text{(similarly for } \int_{-\infty}^{0} \text{)}.$$

On the other hand, the argument of Theorem 2A shows that the hyperbolic terms are $O(T^{\delta})$ for any $\delta > 0$ if $\epsilon = c/\ln T$, $c >> 1$. Indeed all we need is some kind of exponential growth rate for $|\int_\gamma E_s|$ in $L_\gamma$.

For this we claim:

$$|\int_\gamma E(z,\tfrac{1}{2}+ir)| \ll e^{(1/4)L_\gamma}.$$

Indeed, the fundamental domain splits up into the compact part $F_a$ and the ends $\bigcup_\beta \{y_\beta \geq a\}$. We note that $\gamma$ must intersect the compact part since no closed geodesic exists in an end. Let $Y(\gamma) = \max_\beta \max_{z\epsilon\gamma} \{y_\beta(z)\}$; $Y(\gamma)$ is the "excursion height" into an end. Clearly

$$\int_\gamma E(z,\tfrac{1}{2}+ir)| \ll \sum_\beta \int_\gamma y_\beta^{1/2} \ll Y(\gamma)^{1/2}.$$

But $Y(\gamma) \ll e^{\frac{1}{2}L_\gamma}$: for $e^{\frac{1}{2}L_\gamma}$ is the excursion length of a nearly "vertical" geodesic (for a cusp at $\infty$), as the geodesic must emanate from and return to the comapct part. The estimate follows.

It remains to bound $I_{a,i}$ terms (i = 1,2,4). These terms have no analogue in [Z3], but their transcription to the notation of [Z3] is straightforward. Namely, for i = 2,4 (just as for i = 3, handled above), one replaces h(r) in Theorem 3A(iv)-(vi) by $\hat{f}_{T,\epsilon}(r)$. For i = 1, the key factor is written as $M\phi(s/2)$ in Theorem 3A. The $\phi$ appearing there is related to $\hat{f}$ by a Legendre transform ([Za1], (2.24)-(2.25)). We thus have:

$$(4.25.1)\quad I_{a,1}(s,\chi;T,\epsilon) = \frac{1}{4\pi}\int_{-\infty}^{\infty} \hat{f}_{T,\epsilon}(r)\ [\int_0^\infty P_{-\frac{1}{2}+ir}(1+\tfrac{x}{2})\ x^{s/2}\ \tfrac{dx}{x}]\ r\ \mathrm{th}\ \pi r\ dr.$$

The inner integral is equal to:

$$2^s\Gamma(\tfrac{s}{2})\ \frac{\Gamma(\frac{1-s}{2}-ir)\Gamma(\frac{1-s}{2}+ir)}{\Gamma(\frac{1}{2}+ir)\Gamma(\frac{1}{2}-ir)}\quad \text{[G-R, 7.13, 4.2]}.$$

On Re $s = 1/2$ this ratio is asymptotic to a constant times $|r|^{-1/2}$ as $r \to \infty$. So $I_{a,1}(s,\chi;T,\epsilon) \ll \int_{-\infty}^{\infty} \hat{f}_{T,\epsilon}(r)\ |r|^{1/2}\,dr \ll T^{3/4}$ ($\epsilon$ = const./ln T) by an easy argument.

Next,

$$\text{(4.25.2)}\quad \begin{aligned} I_{a,2}(\tfrac{1}{2}+it,s) &= C(s)\int_{-\infty}^{\infty} \frac{\Gamma(\frac{1}{4}+i(r+\frac{t}{2}))\ \Gamma[(\frac{1}{4}+i(\frac{t}{2}-r))]}{\Gamma(ir)\ \Gamma(-ir)}\ \hat{f}_{T,\epsilon}(r)\,dr \\ &\ll \int_{-\infty}^{\infty} |\hat{f}_{T,\epsilon}(r)|\ |r|^{1/2}\,dr \\ &\ll T^{3/2} \end{aligned}$$

by an easy argument. Finally,

$$\text{(4.25.3)}\qquad I_{a,4} \ll \sum_{\substack{\text{exceptional} \\ r_j}} |\hat{f}_{T,\epsilon}(r_j)| \ll 1.$$

This finishes the proof in the weight 0 case.

We next take care of $X_+E_s$. We need, for Re $s = 1/2$,

$$\text{(4.26)}\quad \text{(i)}\ \ \langle Op(X_+E_{a,s})u_j^\theta, u_j^\theta\rangle \ll |r_j|^{1/2}\ln|r_j|$$

$$\text{(ii)}\ \ \int_{\gamma_0} X_+E_{a,s} \ll e^{\frac{1}{4}L_\gamma}$$

(iii) $R_\alpha^+[|E_\beta(\cdot,\frac{1}{2}+ir,\chi)|,s] \ll_s |r|^{1/2} \ln|r| \; [\frac{\Delta'}{\Delta}(\frac{1}{2}+ir)]$

(iv) $I_{\alpha,4}(s,T,\epsilon) \ll 1$

Of course, (ii), (iii), (iv) work exactly as in the weight 0 case. For (ii), (iv) we merely recall that

$$R_\alpha^+[|E_\beta|^2,s] = \frac{A^+(\frac{1}{2}+ir,s)}{\Gamma(\frac{s}{2}+ir)\Gamma(\frac{s}{2}-ir)} R_a[|E_\beta|^2,s].$$

We turn to (i). we explicitly write in the character-dependence of $u_j^\theta$ since (i) = 0 if $\theta = 0$. Then

$$\begin{aligned}\langle Op(E^+E_{\alpha,s})u_j^\theta,\, u_j^\theta\rangle &= C(s)\int_F E(z,s)_{(1)}[L^- u_j^\theta(z)]\overline{u_j^\theta}(z)\, dz \\ &= \int_{F_A} + \int_{y\ge A} = I + II.\end{aligned}$$

$I < A^{1/2}$ as before by the Schwarz inequality, since $\|L^- u_j^\theta\| = 1$.

$\langle Op(E^+E_{\alpha,s})u_j^\theta,\, u_j^\theta\rangle$ splits up similarly, leaving

$$II_+ = \int_{y\ge A} E(z,s)_{(1)}(L^- u_j^\theta)\overline{u_j^\theta}\, dz - \int_{y\ge A} E(z,s)_{(-1)}(L^- u_j^\theta)\overline{u_j^\theta}\, dz$$

(the several cusp case is similar).

Writing $E(z,s)_{(\pm1)} = a_0(y,s)_{\pm1} + G(z,s)_{(\pm1)}$ and noting that $a_0(y,s)_{(+n)} = a_0(y,s)_{(-n)}$, we find that

$$II = \int_{y\ge A} a_0(y,s)_{(1)}[(L^- - L^+)u_j^\theta]\; \overline{u_j^\theta}\, dz$$

$$= \frac{-1}{(s_j+1)} \int_{y\geq A} a_0(y,s)_{(1)} (X_+ u_j^\theta) \, (\overline{u_j^\theta}) \, dz.$$

Here $L^{\pm}u_j^\theta$ and $X_+u_j^\theta$ mean $X_+u_j^\theta(z,\infty)$ qua functions of z. But $X_+u_j^\theta(z,\infty) = y \frac{\partial}{\partial x} u_j^\theta$, so

$$II = \frac{-1}{(s_j+1)} \int_A^\infty a_0(y,s)_{(1)} \{\int_0^1 [\frac{\partial}{\partial x} u_j^\theta(x+iy)] \, \overline{u_j^\theta}(x+iy) \, dx\} \frac{dy}{y}.$$

When $\theta = 0$, the dx integral vanishes (the trivial case). If $\theta \neq 0$ one goes back to imitate the estimate of the weight 0 case, with now $u_j^\theta(x+iy) = \sum_{n\neq 0} \rho_j^\theta(n) \, y^{1/2} K_{ir_j}(2\pi|n|y) \, e(nx)$.

The derivative picks up an additional order in n, hence $s_j$, which is cancelled by the factor $\frac{1}{s_j+1}$. So the estimate goes through as before.

We conclude:

$$\sum_{r_j\leq T} \langle Op(E^+E_{\alpha,s})u_j^\theta, \, u_j^\theta\rangle - \frac{1}{4\pi}\int_{-T}^{T} R_\alpha^+[|E_\beta(\cdot,\tfrac{1}{2}+ir,\chi)|^2,s] \, dr$$
$$\ll T^{3/2} \text{ on } \mathrm{Re}\ s = 1/2.$$

Finally we must deal with the residual terms. However the only facts we used about Eisenstein series related to their Fouier expansions, in particular the growth rate of the zero$^{th}$ order terms in cusps. The coefficients $\phi_{\alpha\beta n}(s)$ are the only things that changes in the residual terms, and consequently only the (irrelevant) s-dependence of estimates changes.

This concludes the proof of Theorem 4B. ∎

Finally, we give the proof of Proposition 4C.

Proof. (a) For cuspidal $\sigma$, the proposition is simple. In the weight 0 case, it follow immediately from the estimates $|\langle u_k u_n, u_n\rangle| \underset{k}{<<} 1$ and (4.10). In the higher weight case, it follows likewise from $\|Op(\sigma)\| << 1$ and (4.16).

(b) For Eisenstein $\sigma$, on the other hand, the estimates (4.21), (4.23), (4.26)(i) are not good enough to immediately imply the proposition. However, a truncation argument combined with Theorem 4B does imply it. We illustrate the argument in the weght 0 case; the higher weight ones are reasonably straightforward modifications.

So we let $\sigma = E(\cdot,s)$ (Re $s = 1/2$). Let us consider first

$$|N|(\sigma,\chi,T) = \sum_{|r_j|\leq T} |\langle E_s u_j, u_j\rangle|.$$

It is clear that for any A (with $y \geq A$ a cuspidal end)

$$(4.27) \qquad |N|(\sigma,\chi,T) \leq \sum_{|r_j|\leq T} |\langle \tilde{E}_{A,s} u_j, u_j\rangle| + \sum_{|r_j|\leq T} |\langle (1-\chi_A) a_0(\cdot,s) u_j, u_j\rangle|$$

($\chi_A$ being the characteristic function of $F_A = F \cap \{y \leq A\}$).

Since $\tilde{E}_{A,s}$ is bounded, the first term is $O(\|\tilde{E}_{A,s}\|)\, N_\Gamma(T)$. The inner product in the second term is $O_s(\int_{y\geq A} y^{1/2}|u_j|^2\, dz)$ ($s$ on Re $s = 1/2$). Consider $\dot{E}(z,\frac{1}{2}) = \frac{d}{ds}\Big|_{s=1/2} E(Z,s)$. The zero$^{th}$ Fourier coefficient of $\dot{E}(z,\frac{1}{2})$ is $2y^{1/2} \ln y + \phi'(\frac{1}{2})y^{1/2} = \dot{a}(y,\frac{1}{2})$. Hence

$$\int_{y\geq A} y^{1/2}|u_j|^2\, dz \leq |\int_F \dot{E}(z,\tfrac{1}{2})|u_j(z)|^2\, dz| + O(1) \text{ as } j \to \infty.$$

It follows that

$$(4.28) \qquad \sum_{|r_j|\le T} |\langle(1-\chi_A)a_0(\cdot,s)u_j,u_j\rangle| \le \sum_{|r_j|\le T} |\langle\hat{E}(\cdot,\tfrac{1}{2})u_j,u_j\rangle| + O(N_\Gamma(T)).$$

(Note. We use $\hat{E}(z,\frac{1}{2})$ because $E(z,\frac{1}{2})$ may be cuspidal; compare [Z5], §6.)

Hence:

$$(4.29) \qquad |N|(\sigma,\chi,T) \ll \sum_{|r_j|\le T} |\langle\hat{E}(\cdot,\tfrac{1}{2})u_j,u_j\rangle| + O(N_\Gamma(T))$$

Now consider $|M|(\sigma,\chi,T)$.

It will be helpful to use the notation:

$$(4.30) \qquad R_\alpha(|E_\beta(\cdot,\tfrac{1}{2}+ir,\chi)|^2,s) = RN[\ E_\alpha(\cdot,s)E_\beta(\cdot,\tfrac{1}{2}+ir,\chi),E_\beta(\cdot,\tfrac{1}{2}+ir,\chi)\ ],$$

where the renormalized inner product is defined in [Za2] (see also [Z5]). We recall that a "renormalizable" function is a function $H$ satisfying (in any cusp)

$$(4.31) \qquad H = h + O(y^{-N}) \quad (\forall\ N)$$

where $h$ is a sum of homogeneous functions (including logarithmic factors), and that

$$(4.32) \qquad RN(\int_{\Gamma\backslash h} H\,dz) = \int_{F_T} H + \int_{F-F_T} (H-h) - \hat{h}(T),$$

where $\hat{h}$ is an anti-derivative of $y^{-2}h$. (See [Za1] or [Z5] for a more complete discussion.)

A similar definition can be given by $RN[Op(\sigma)E_\beta(\cdot,\frac{1}{2}+ir,\chi),E_\beta(\cdot,\frac{1}{2}+ir,\chi)]$; of course, it defines a linear functional of (renormalizable) $\sigma$.

In these terms, our task is to estimate

$$|M|(\sigma,\chi,T) = \int_{-T}^{T} |RN \langle Op(\sigma)E(\cdot,\tfrac{1}{2}+ir),E(\cdot,\tfrac{1}{2}+ir)\rangle|\, dr, \tag{4.32a}$$

where from here on we again pretend $\chi \equiv 1$ and that $\Gamma$ has one cusp at $\infty$. Furthermore, we will again only give details on the weight 0 case, where $\sigma = E_s$ ($\mathrm{Re}\, s = 1/2$); the other cases are quite similar.

As in (4.27), we will estimate separately the terms

$$\text{(4.33) (i)} \qquad \int_{-T}^{T} |RN \langle \tilde{E}_{A,s}E(\cdot,\tfrac{1}{2}+ir),E(\cdot,\tfrac{1}{2}+ir)\rangle|\, dr$$

$$\text{(ii)} \qquad \int_{-T}^{T} |RN \langle (1-\chi_A)a_0(\cdot,s)E(\cdot,\tfrac{1}{2}+ir),E(\cdot,\tfrac{1}{2}+ir)\rangle|\, dr$$

Since $\tilde{E}_{A,s}$ is cuspidal in an end, it is straightforward to adapt the estimate of (4.12). We get:

$$\text{Term (4.33) (i)} = O_{s,A}\left(\left|\tfrac{\Delta'}{\Delta}(\tfrac{1}{2}+ir)\right|\right). \tag{4.34}$$

Here the (s,A) dependence plays no significant role.

The term (4.33) (ii) may be easily computed. Setting $A = T$ in (4.33),

$$\text{Term (4.33) (ii)} = \int_{-T}^{T} \left| \int_{A}^{\infty} a_0(y,s)\theta^0(|G(.,\tfrac{1}{2}+ir)|^2)\, \frac{dy}{y^2} \right| dr. \tag{4.35}$$

(Compare (3.8.5); the $h_A(s)$ terms there make lower order contributions to (4.33) (ii).) It follows that

(4.36) $$\text{Term (4.33 (ii)} \ll_s \int_{-T}^{T}\left[\int_A^\infty y^{1/2}\theta^0|G(.,\tfrac{1}{2}+ir)|^2(y)\,\frac{dy}{y^2}\right] dr$$
$$\ll \int_{-T}^{T} dr\left[\int_A^\infty \dot{a}(y,\tfrac{1}{2})\theta^0|G(.,\tfrac{1}{2}+ir)|^2(y)\,\frac{dy}{y^2}\right]$$
$$= \int_{-T}^{T} RN\, \langle\dot{E}(\cdot,\tfrac{1}{2})E(\cdot,\tfrac{1}{2}+ir),E(\cdot,\tfrac{1}{2}+ir)\rangle\, dr + O_A(|\tfrac{\Delta'}{\Delta}(\tfrac{1}{2}+ir)|).$$

Combining (4.29) and (4.36), we have:

(4.37) $$(|N| + |M|)(\sigma,\chi,T) \ll (N + M)(\dot{E}(\cdot,\tfrac{1}{2}),T) + O(N_\Gamma + M_\Gamma)(T).$$

But the proof of Theorem 4B applies with no essential change to $\sigma = \dot{E}(\cdot,\frac{1}{2})$, except that the zero$^{th}$ Fourier coefficient acquires a logarithmic factor. This changes $A^{-1/2}$ in the proof of (4.20) to $\ln A\cdot A^{-1/2}$, so the estimate of (4.21) acquires a new factor of $\ln|r_j|$. Hence,

(4.38) $$(N + M)(\dot{E}(\cdot,\tfrac{1}{2}),T) \ll T^{3/2}\ln T.$$

We conclude:

(4.39) $$(|N| + |M|)(E_s,\chi,T) \ll (N_\Gamma + M_\Gamma)(T),$$

as desired. ∎

## § 5. Asymptotic Expansions

In this section we establish the prime geodesic theorems: that is, the asymptotic expansions as $T \to \infty$ of the sum functions $\Psi_\Gamma(\sigma,T) = \sum_{L_\gamma \le T} \int_\gamma \sigma$.

These expansions are derived from the trace formulae of §2-§3, along the lines indicated in §1. The arguments are essentially the same as in [Z2] for the compact case, modulo the new technical complications caused by continuous spectral terms. For the most part, these technical problems have been handled in §4. Hence, the main thrust of this section is to explain how to combine the results of §2-4 of this paper with the estimates of [Z2, §4]. The latter material will not be reviewed here (see §1 and the Appendix). On the whole, the reader will be asked to refer to [Z2] for relevant prior results.

One rather technical matter must be addressed before we can begin the asymptotic analysis. Namely, we must confirm the validity of the trace formulae for the test functions $\phi_{T,\epsilon}$, that will be used below in deriving the prime geodesic theorems. As in [Z2], §4, the $\phi_{T,\epsilon}$ are defined so that (non-standard) HC transforms, $H_{(m,s)}\phi_{T,\epsilon}$, approximate characteristic functions of length intervals. The functions $H_{(m,s)}\phi_{T,\epsilon}$ will belong to $C_0^\infty$; however, the $\phi_{T,\epsilon}$ themselves will not . Nevertheless, the various transforms of $\phi_{T,\epsilon}$ which appear in the trace formulae have rapid enough decay that the formulae remain valid for them anyway. The proof of this is a straightforward continuity argument. We begin this section by giving the non-routine details of it. After that, we will explain how to modify the estimates of [Z2], §4 so that they yield asymptotic expansions for the general finite area case.

The trace formulae of this paper may be written in a simplified notation as:

(5.1) $$\mathrm{Spec}(\phi) = \mathrm{hyp}(\phi) + \mathrm{ell}(\phi) + \mathrm{res}(\phi) + \mathrm{standard}(\phi),$$

where:

(5.1a) $$\mathrm{Spec}(\phi) = \sum_j (\mathrm{Op}(\sigma)u_j,u_j)S_m\phi(ir_j) + \int \mathrm{RN}(\mathrm{Op}(\sigma)E_{ir},E_{ir})S_m\phi(ir)\, dr;$$

(5.1b) $$\mathrm{hyp}(\phi) = \sum_{\{\gamma\}_{hyp}} (\int_{\gamma_0} \sigma)H_{s,m}\phi(L_\gamma);$$

where the elliptic and residual terms are finite sums of the same kinds (see §2-§3); and where standard($\phi$) are the extra terms occuring in §3, denoted "standard" here because they involve expressions essentially independent of the discrete group $\Gamma$.

The formulae (5.1) have been proved in §2-§3 for $\phi \in C_0^\infty$. To extend them to a wider class of $\phi$, we must first give some straightforward bounds on the terms in (5.1), viewed as functionals of $\phi$.

The main terms may be written (see Appendix, §1 and §4 for notation):

(5.2a) $$\mathrm{spec}(\sigma) = \int_0^\infty S_m\phi(ir)d(M+N)(\sigma,r)$$

(5.2b) $$\mathrm{hyp}(\sigma) = \int_0^\infty H_{(s,m)}\phi(\ell)d\psi_\Gamma(\sigma,\ell) .$$

The functionals spec and hyp can be bounded as follows (for any $\gamma > 0$):

(5.3a) $$|\mathrm{spec}(\phi)| \leq C(\gamma,s,m) \sup_{r\in\mathbb{R}} |S_m\phi(ir)(1+r^2)^{(1+\gamma)}|$$

(5.3b)
$$|\mathrm{hyp}(\phi)| \leq C(\gamma,s,m) \operatorname*{sum}_{\ell\in\mathbb{R}^+} \left| H|\phi|(\ell) e^{5/4\ell} (1+\ell^2)^{(\frac{1}{2}+\gamma)} \right|$$

Estimate (5.3a) follows immediately from

(5.4)
$$\int (1+r^2)^{-(1+\gamma)} d(|M|+|N|)(\sigma,r) < \infty ,$$

which in turn follows from §4 [Proposition C]. Estimate (5.4) follows from the pair:

(5.5a)
$$|H_{m,s}\varphi(\ell)| \leq C(s,m) H(|\varphi|)(\ell)$$

(5.5b)
$$\int_0^\infty (1+\ell^2)^{-(\frac{1}{2}+\gamma)} e^{-\frac{5}{4}\ell} d|\Psi_\Gamma|(\sigma,\ell) < \infty .$$

The first of these two is just the statement that the hypergeometric functions appearing in the formulae for $H_{m,s}$ are bounded on $\mathbb{R}$ (see [Z2], §2). This boundedness follows from the so-called connection formulae, relating the values of the hypergeometric function at 0 and $\infty$ ([G-R, 9.132). The second statement follows from the standard prime geodesic theorem, and the fact that $|\int_\gamma \sigma| \leq C(s,m) e^{\frac{1}{4}l(\gamma)}$ for the $\sigma$ of §2-§3. Indeed, the periods are bounded in $L(\gamma)$ except when $\sigma$ is an Eisenstein series $E(\cdot,\frac{1}{2}+ir)$, or a derivative or residue of same; and then the bound is clear (see the proof of Theorem 4B).

The remaining terms in the trace formulae (5.1) can easily be bounded by the norms in (5.3). To save space, we leave the details to the reader.

All functionals in the trace formulae are therefore bounded by the norms ($\gamma > 0$):

(5.6)
$$\|\phi\|_{(\gamma)} = \sup_r |S_m\phi(ir)(1+r^2)^{(1+\gamma)}|$$
$$+ \sup_{\ell\in\mathbb{R}^+} \left| H(|\phi|)(\ell)e^{5/4\ell}(1+\ell^2)^{(\frac{1}{2}+\gamma)} \right| .$$

The identity expressed by the trace formulae between these functionals on $C_0^\infty$ must therefore continue to hold on the closure of $C_0^\infty$ under (some of) these norms. It remains for us to check that the $\phi_{T,\epsilon}$ to be used below for the prime geodesic theorems all lie in this closure.

The $\phi_{T,\epsilon}$ are identical to the ones used in [Z2] §4 in the compact case. They were constructed there to satisfy:

(5.7a)
$$H_{(m,s)}\phi_{T,\epsilon} \in C_0^\infty$$

(5.7b)
$$\|\phi_{T,\epsilon}\|_{(\gamma)} < \infty \text{ for all } 0 < \gamma < \tfrac{1}{2} .$$

Infact they have the further property:

(5.7c)
$$\phi_{T,\epsilon} \in C_c(G) .$$

Assuming (5.7a-c), it follows by a familiar smoothing argument that the $\phi_{T,\epsilon}$ lie in the closure of $C_0^\infty$. Indeed, let $\rho_\delta \in C_0^\infty \cap S_{0,0}$ be an approximate identity. Suppose that $\phi \in S_{m,0} \cap C_c(G)$, and that $\|\phi\|_\gamma < \infty$. We claim:

$$(5.8) \qquad \|\rho_\delta*\phi-\phi\|_{(\gamma')} \longrightarrow 0 \text{ as } \delta \longrightarrow 0 \text{ for all } 0 < \gamma' < \gamma .$$

To see this, we first note that for $\phi \in C_c \cap S_{m,0}$:

$$(5.9) \qquad S_m(\rho_\delta*\phi)(s) = S\rho(\delta s)S_m\phi(s) .$$

Indeed, by [Z2] (1.14), $S_m = MH_m$, where $M$ is the usual Mellin transform on $\mathbb{R}^+ \simeq A^+$ and $H_m$ is the transform on $C_c \cap S_{m,0}$ given by:

$$(5.10) \qquad H_m\phi(a) = \rho(a)\int_N \phi(an)dn$$

(see [L], p. 69 for notation). Then (5.9) comes down to the following convolution formula along $A$:

$$(5.11) \qquad H_m(\rho*\phi) = H\rho*H_m\phi \quad (\rho \in C_c \cap S_{0,0};\ \phi \in C_c\cap S_{m,0}) .$$

The proof of (5.11) is essentially the same as for $m = 0$ ([L], pp.70-71).

Now suppose $\|\phi\|_\gamma < \infty$, fix $\gamma' < \gamma$ and let $\epsilon' = \gamma-\gamma'$. Then the first term of $\|\rho_\delta*\phi-\phi\|$ is:

$$(5.12) \qquad \sup_r |(S_m(\rho_\delta*\phi)-\phi(ir))(1+r^2)^{(1+\gamma')}|$$

$$\leq \sup_r |S_m\phi(ir)(1+r^2)^{(1+\gamma)}| \sup_r |(S\rho(\delta ir)-1)(1+r^2)^{-\epsilon'}|$$

$$\leq C(\rho,\epsilon_0')\|\phi\|_\gamma \sup_r |[M(H\rho)(\delta ir)-1]\cdot(1+r^2)^{-\epsilon_0'}| .$$

Since $H\rho \in C_0^\infty(A)$, with $\mathbb{M}H\rho(0) = 1$, one knows that, as $\delta \longrightarrow 0$, $\mathbb{M}(H\rho)(\delta ir)-1$ converges to zero uniformly on compact sets of $\mathbb{R}$. It follows that

$\sup_r |(\mathbb{M}(H\rho)(\delta ir-1))(1+r^2)^{-\epsilon'}|$ tends to zero with $\delta$.

This leaves the second term of $\|\rho_\delta*\phi-\phi\|_{(\gamma')}$:

$$(5.13) \qquad \sup_\ell \left| H(|\rho_\delta*\phi-\phi|(\ell)e^{5/4\ell}(1+\ell^2)^{(\frac{1}{2}+\gamma)}\right| .$$

For $\phi \in C_c \cap S_{m,0}$ it is quite easy to show this tends to zero with $\delta$. Indeed, for $\delta < 1$, $\rho_\delta*\phi-\phi$ is supported in some fixed compact $\Omega \subset G$. It follows that $H(|\rho_\delta*\phi-\phi|)$ is supported in a fixed compact $\Omega_+$ of $A_+$ (or of $\mathbb{R}^+$ if we identify $\begin{bmatrix} e^{-\ell/2} & 0 \\ 0 & e^{\ell/2} \end{bmatrix}$ with $\ell$). The right factor of (5.13) is bounded on $\Omega_+$, so we just need that $H(|\rho_\delta*\phi-\phi|)$ tends uniformly to zero in this set. Since hyperbolic conjugacy classes intersect any compact $\Omega \subset G$ in sets of uniformly bounded measure, $|H(\rho_\delta*\phi-\phi)(\ell)| \le C(\Omega)\|\rho_\delta*\phi-\phi\|_\infty$. But $\phi \in C_c(G)$ implies $\rho_\delta*\phi \longrightarrow \phi$ uniformly.

This concludes the proof of (5.8).

It now remains to show that (5.7a-c) hold for the $\phi_{T,\epsilon}$ of [Z2] §4. As mentioned above, (5.7a-b) are contained in that paper. As for (5.7c), we now explain the key point in the weight 0 case; the other cases are very similar.

The only fact about $\phi_{T,\epsilon}$ we will need to use for (5.7c) is the fact ((5.7a)) that $H_{s,m}\phi_{T,\epsilon}$ is smooth and compactly supported. Following [Z2] §3, we change variables to $v = (\mathrm{sh}\frac{\ell}{2})^2$. Evidently $H_{s,m}\phi_{T,\epsilon} \in C_0^\infty(\mathbb{R}_v^+)$. Now, let $\psi$ be any element of $C_0^\infty(\mathbb{R}_v^+)$. From [Z2] (Proposition 3.1), we have a explicit formula for $H_{s,0}^{-1}\psi$. To keep the notation consistent with that paper, we write

$H_k^c$ for $H_{s,0}$ ($s_k$ parametrize the principal series representations in $\Gamma|G$). The formula is:

$$(5.14) \qquad (H_k^c)^{-1}\psi(v) = \int_{\mathrm{Re}\ s=s_0} M\psi(s)\mu_k^c(s)^{-1}v^{-s+\frac{1}{2}}\frac{ds}{i} \quad (s_0 > \tfrac{1}{2})$$

where $M\psi$ is the Mellin transform on $\mathbb{R}_v^+$, and where

$$(5.14a) \qquad \mu_k^c(s)^{-1} = \frac{\Gamma(s)^2}{\Gamma(\frac{1}{2})\Gamma(-\frac{1}{4}+\frac{ir_k}{2}+s)\Gamma(-\frac{1}{2}-\frac{ir_k}{2}+s)} .$$

Since $M\psi(s)$ is in the Paley-Wiener space $PW(\mathbb{C})$ (A15), and since $\mu_k^c(s)^{-1}$ has no poles in Re $s_0 > \frac{1}{2}$ , the line of integration may be shifted rightwards. By a familiar argument ([L], p.77), the integral must vanish for sufficiently large v. Recalling that $v = \frac{1}{2}(e^{\ell/2}-e^{-\ell/2})$, it follows that $(H_k^c)^{-1}\psi$, interpreted as a function on $A^+$, is compactly supported there. The same obviously holds for the extension to G as an element of $S_{m,0}$. Continuity of $(H_k^c)^{-1}\psi$ is clear from the expression (5.14). (5.7c) is thus confirmed.

Although it will not be needed below, it may be helpful to explain why $(H_k^c)^{-1}\psi$ is not smooth. The problem is that if we shift the contour leftwards, we can pick up residues at $s = 0,-1,\cdots$. So $(H_k^c)^{-1}\psi$ has an expansion in half-integral powers of v as $v \longrightarrow 0$. Since $v^{1/2} = |\mathrm{sh}\frac{\ell}{2}|$, we can get singularities at $\ell = 0$ (i.e. at the identity element). The conditions (such as $M\psi(0) = 0$) that we placed on $\psi$ in ([Z2] §4) removed the worst singularity, and as a result (5.7a-b) hold. Finally, we note that the singularity could probably have been removed by normalizing the transform differently (i.e. by multiplying the the factor $|D(a)|$ as in A16(b)). However, this would just change the point in the proof where the singularity occurred.

Having confirmed the validity of the trace formulae of §2-§3 for the $\phi_{T,\epsilon}$ of ([Z2] §4), we now proceed to the proof of the prime geodesic theorems. We will only give full details on new aspects not already analyzed in [Z2].

Cuspidal Case. For $\sigma \in \{u_k, X_+u_k, \psi_m\}$ a cusp form, we have:

Theorem 5A: $$\Psi_\Gamma(\sigma,\chi_\theta,T) = \sum_{j=1}^{M} \langle Op(\sigma)u_j^\theta, u_j^\theta\rangle \gamma_{s,j}^\theta e^{(\frac{1}{2}+t_j)T} + O(Te^{\frac{3}{4}T}) ,$$

where $\Omega\sigma = s(1-s)\sigma$, where $t_j = |r_j|$, where the sum extends over the complementary series representations and where $\gamma_{s,j}^\theta$ are exactly the constants appearing in [Z2] §4 in the analogous expansions. For example, if $\sigma = u_k$, $\theta = 0$, and $s_k = \frac{1}{2}+ir_k$, then

$$\gamma_{s_k,j}^\theta = \frac{\Gamma(\frac{1}{2}+t_j)\Gamma(t_j)\Gamma(1-2t_j)}{\Gamma(t_j+\frac{1}{4}+\frac{1}{2}ir_k)\Gamma(t_j+\frac{1}{4}\, ir_k)(\frac{1}{2}+t_j)}$$

(Note: The $s_k$ of this paper differ from that of [Z1]; see §1. Also, in the $X_+u_k$ case it might be cleaner to write $\gamma_{s_k,j}^{\theta,+}$, say, since these constants are not those of the $u_k$ case).

Proof (compare [Z2] §4, Theorems 4.1, 4.2, 4.3). For cuspidal $\sigma$, the trace formulae of §2 acquire, in addition to the hyperbolic and discrete spectral terms in [Z2] §2, the new elliptic and continuous spectral terms. Except when $\sigma$ has weight 0, the elliptic terms vanish.

We substitute into the trace formula the function $\phi_{T,\epsilon}$ whose HC transforms equal the $\psi_{T,\epsilon}$ of [Z2] (loc. cit.). These $\psi_{T,\epsilon}$ are the smoothed out characteristic functions discussed above.

As in [Z2] §4, we then have:

$$\Psi_\Gamma(\sigma,\chi,T) = (C)_{T,\epsilon} + (P)_{T,\epsilon} + (ell)_{T,\epsilon} + O(\epsilon e^T) .$$

Here (C), (P), (ell) stand for complementary series, principal series and elliptic terms respectively. The error $O(\epsilon e^T)$ comes from estimating the hyperbolic terms, exactly as in [Z2].

Just as in the proof of the standard prime geodesic theorem, the elliptic terms are bounded independently of $(t,\epsilon)$. The proof is straightforward (or, see [C-S]).

The only substantial change from [Z2] is the $(P)_{T,\epsilon}$ now includes the continuous spectral term

$$\int_{-\infty}^{\infty} (S_m H_{s,m}^{-1}) \psi_{T,\epsilon}(ir) dM(\sigma,r) dr .$$

This has just the effect of replacing occurrences of $N(\sigma,t)$ in the proofs of the above-cited theorems in the compact case by $(N+M)(\sigma,t)$. However, by §4, Theorem A, (N+M) satisfies the same $O(t/\log t)$ growth estimate as did N in the compact case. This growth estimate is all that we used about N in [Z2] §4. Hence, all estimates there generalize to the co-finite case.

<u>Eisenstein and residual symbols</u> $\sigma \in \{E_{a,s};\ X_+ E_{a,s};\ \text{residue}\}$. The only change in the Eisenstein case is in the error estimate. We have:

<u>Theorem</u> <u>5B</u>. For Eisenstein $\sigma$, with Re s $= \frac{1}{2}$,

$$\Psi_\Gamma(\sigma,\chi_\theta,T) = \sum_{j=1}^{M} \langle \mathrm{Op}(\sigma)u_j^\theta, u_j^\theta\rangle \gamma_{s,j}^\theta e^{(\frac{1}{2}+t_j)T} + O(e^{19/20T}) ,$$

where all constants are exactly as in the $\{u_k, X_+u_k\}$ cuspidal cases.

<u>Proof</u>. The estimates of [Z2] §4 (Theorems 4.1 and 4.2) now change in three ways.

First, the estimate on the hyperbolic terms must be changed. We write (dropping the $\chi$):

$$\begin{aligned} \Psi_\Gamma(\sigma,T) &= \sum_{L_\gamma \le T} \int_\gamma \sigma \\ &= \sum_{\{\gamma\}} (\int_\gamma \sigma)\, \psi_T(L_\gamma) \\ &= \sum_{\{\gamma\}} (\int_\gamma \sigma)\, \psi_{T,\epsilon}(L_\gamma) + O\Big[\sum_{T-\epsilon \le L_\gamma \le T+\epsilon} |\int_\gamma \sigma|\Big] . \end{aligned}$$

Here $\psi_T$ is essentially a characteristic function; and $\psi_{T,\epsilon}$ a smoothing of it ([Z2] §4). Instead of the estimates $|\int_\gamma \sigma| = O(1)$ as $L(\gamma) \longrightarrow \infty$, which hold in the compact or cuspidal cases, we now have only: $|\int_\gamma \sigma| = O(e^{\frac{1}{4}L(\gamma)})$. Accordingly the error estimate $O(\epsilon e^T)$ for the second term has to be changed to $O(\epsilon e^{5/4T})$.

The second change is in the principal series terms $(P)_{T,\epsilon}$. As in the cuspidal case, occurrences of $N(\sigma,r)$ in [Z2] §4 must be replaced by

$(N+M)(\sigma,T)$. However, by §4, Theorem 4B, we now have: $(M+N)(\sigma,T) << T^{3/2}$. The key point where this affects the proof of [Z2], Theorem 4.1, is in the estimate of the main error term:

$$\mathrm{Res}_{1/2}(P)_{T,\epsilon} = Te^{T/2}\int_0^\infty |N(u_k,t)| \left|\frac{1}{|\frac{1}{2}\pm it|}\right| |M\phi(\epsilon(\tfrac{1}{2}\pm it)|dt .$$

We now replace $|N|$ by $|M+N|$. The $\epsilon$-dependence of $\mathrm{Res}_{1/2}(P)_{T,\epsilon}$ then changes from $\epsilon^{-1}$ in [Z1] to $\epsilon^{-3/2}$ here.

So far, this changes the expansion of $\Phi_\Gamma(\sigma,T)$ (up to inessential factors of T) from

$$\Phi_\Gamma(\sigma,T) = (C)_{T,\epsilon} + O(\epsilon e^T) + e^{T/2}O(\epsilon^{-1})$$

in the campact case to

$$\Phi_\Gamma(\sigma,T) = (C)_{T,\epsilon} + O(\epsilon e^{5/4T}) + O(\epsilon^{-3/2}e^{T/2}) + (\mathrm{standard})_{T,\epsilon}$$

in the Eisenstein case. Here (standard) represents the terms $I_{\alpha,j}(s,\chi)H_k^{-1}\phi_{T,\epsilon}$ ($j=1,2,4$) from the weight 0 case; and similarly for the case of $X_+E_s$. The third and final change from [Z2] is to estimate the contribution of these terms. The only non-trivial one among them is where $j = 2$. This has the same form as the $M$ term ($j=3$), except we are integrating $H_k^{-1}\phi_{T,\epsilon}$ ($= h_{T,\epsilon}$) against a quotient of $\Gamma$-values which only grows like $r^{1/2}$. So the standard terms can be easily absorbed into $(P)_{T,\epsilon}$.

The error term in the asymptotic expansion is thus of the form $O(\epsilon e^{5/4T})$ $+ O(\epsilon^{-3/2} e^{T/2})$. The optimal $\epsilon$ is $\epsilon = e^{-3/10T}$, leaving the error stated above.

∎

§ 6. Equidistribution of closed geodesics

From the asymptotic expansions of §5, it is a short step to prove that closed geodesics tend to become uniformly distributed relative to Haar measure as the period tends to $\infty$.

First, we should clarify the meaning of uniform distribution.

Let us set:

(i) $\mu_\gamma$ = periodic orbit measure: $\mu_\gamma(f) = \int_\gamma f$,

(ii) $\mu_T = (\sum_{L_\gamma \leq T} L_\gamma)^{-1} \sum_{L_\gamma \leq T} \mu_\gamma$,

(iii) $\mu$ = normalized Haar measure: $\mu(\Gamma\backslash h) = 1$

(iv) $C_0(\Gamma\backslash G)$ = continuous functions vanishing at $\infty$,

(v) $C_0^\infty$ = smooth, compactly supported functions

(vi) $C_b$ = bounded continuous functions.

We will say that closed geodesic tend on average to be uniformly distributed relative to $\mu$ if $\mu_T$ tends to $\mu$ weakly on $C_0$ (i.e. $\mu_T(f) \longrightarrow \mu(f)$ for $f \in C_0$). As we will show below, this automatically implies $\mu_T \longrightarrow \mu$ weakly on $C_b$.

Geometrically, it seems more natural to ask whether closed geodesics tend generically to be uniformly distributed relative to $\mu$. By definition, this will mean:

$$(\sum_{L_\gamma \leq T} L_\gamma)^{-1} \sum_{L_\gamma \leq T} |\mu_\gamma - \mu(f)| \longrightarrow 0$$

as $T \longrightarrow \infty$ for $f \in C_0$.

Generic uniform distribution implies that $\frac{1}{L_\gamma} \mu_\gamma(f)$ tends to $\mu(f)$ along a sequence of $\gamma$'s of (counting) density one in the length spectrum.

Combining the facts that the geodesic flow $G^t$ is ergodic on finite area quotients $\Gamma\backslash h$ and that the $\mu_\gamma$ are positive invariant measure for $G^t$, one can show that <u>on average</u> uniform distribution implies <u>generic</u> uniform distribution for these spaces. The argument is exactly as for the compact case (cf. [Z2] §5). So henceforth we will only consider weak convergence of $\mu_T$ to $\mu$.

We begin by collecting the results of §5, together with some of [Z2]. Recall that the lowest non-zero eigenvalue $\lambda_1$ of $\Gamma\backslash h$ is of the from $\lambda_1 = \frac{1}{4} - t_1^2$ where $0 \leq t_1 < \frac{1}{2}$.

<u>Theorem 6.1</u>: Let $\sigma$ be an eigenform of weight m of the Casimir: $\Omega\sigma = s(1-s)\sigma$, $\frac{1}{i} W\sigma = m\sigma$. Assume $\frac{1}{2} \leq \operatorname{Re} s < 1$, and that $\sigma \perp 1$. Then:

$$\mu_T(\sigma) = O_{s,m}(e^{(t_1-1/2)T}) + O_{s,m}(e^{-1/20T}).$$

<u>Proof</u>. For $\sigma$ among $\{u_k, X_+u_k, \psi_m, E_s, X_tE_s\}$, the estimate follows immediately from the expansions of §5, together with the prime geodesic theorem $(\mu_T(1) \sim ce^T)$. These spectral forms give a cyclic basis for the action of the geodesic flow (i.e. the right action of A on $\Gamma\backslash G$). Since s lies in one irreducible component of $\Gamma\backslash G$ and has a weight, the period $\mu_\gamma(\sigma)$ can be expressed as a (finite) linear combination of the periods of the cyclic basis elements of its irreducible (see [Z2] §1). So the same decay as $T \longrightarrow \infty$ holds for it. ∎

In this theorem we have not estimated the dependence of the O-symbols on s. This is for a good reason. Indeed, even the dependence of the asymptote on s is at present beyond our means to estimate accurately. The problem is with the coefficients $\langle Op(\sigma)u_j, u_j\rangle \gamma_{s,j}$ (cf. Theorem 5A). The explicit expression for $\gamma_{s,j}$ (loc. cit.) combined with Stirling's formula show that $\gamma_{s,j}$ increases like $|s|^A e^{\frac{\pi}{2}|s|}$ as $|s| \to \infty$. It is possible that the matrix element $\langle Op(\sigma)u_j, u_j\rangle$ cancels the exponential growth rate, but this is at present unknown. Hence the best estimate we could give at present in the s-dependence of the O-symbol has the form $|s|^A e^{\frac{\pi}{2}|s|}$. This estimate also takes into account the dependence on the periods $\int_\gamma \sigma$ and on the hypergeometric functions which appear in the transforms. Since it would require a lengthy digression to prove this estimate completely, we would prefer to avoid having to use it. The only statement we will use is that the dependence of the O-symbols on $|s|$ is continuous. This is so straightforward that details may be left to the reader. We then can prove weak convergence of $\mu_T \to \mu$ on $C_b$ by a simple density argument.

To proceed with this argument, we first invoke that $C_0^\infty(\Gamma\backslash G)$ is uniformly dense on $C_0(\Gamma\backslash G)$: the proof is obvious. So let $f \in C_0^\infty$, and split up f into its cuspidal- and $\theta$-components: $f = {}^0f + {}^\theta f_0$, where

(6.2i) $$ {}^0f = \sum_{m,j} \hat{f}(s_j, m)\, u_{j,m} $$

(6.2ii) $$ {}^\theta f = {}^{res}f + \sum_m \int_{\mathrm{Re}\, s=1/2} \hat{f}(s_j, m) E(s,\cdot)_{(m)}\, ds $$

(6.2iii) $$ \hat{f}(s_j, m) = \langle f, u_{j,m}\rangle; \quad \hat{f}(s,m) = \langle f, E(s,\cdot)_{(m)}\rangle $$

Note that Theorem 6.1 already applies to the residual part ${}^{res}f$ of f; so henceforth, we can pretend it doesn't exist.

Now, from $f \in C_0^\infty$ we get, by partial integration, that $|\hat{f}(s_j,m)| \leq (1 + |s_j|^2 + m^2)^{-N} \langle (1-\Delta)^N f, u_{j,m}\rangle$, where $\Delta$ is the elliptic Laplacian $\Omega + 2W^2$. The Schwarz inequality then implies $|\hat{f}(s_j,m)| \ll (1 + |s_j| + |m|)^{-N}$ (all N). The same kind of argument also applies to the coefficients $\hat{f}(s,m)$ since $f \in C_0^\infty$. Indeed, we view $\hat{f}(s,m) = (1 + |s|^2 + |m|^2)^{-N} \langle (1-\Delta)^N f, E(\cdot,s)_{(m)}\rangle$ as an integral over the support of f, which is contained in some $F_A$. Then the Schwarz inequlity impies that

$$|\hat{f}(s,m)| \leq C(f)(1 + |s| + |m|)^{-N} \left(\int_{F_A} |E(\cdot,s)_{(m)}|^2 \, dz\right)^{1/2}$$
$$\leq C(f)(1 + |s| + |m|)^{-N} \left|\frac{\Delta'}{\Delta}(s)\right|^{1/2} \qquad (\mathrm{Re}\ s = \tfrac{1}{2}).$$

In the last line $\Delta(s) = \det \Phi(s)$ (A10), and the inequality on the integral comes from the Maass-Selberg inner product formula (see, e.g. [Sa] §3).

The main step in the density argument is then:

<u>Proposition 6.3</u>. (i) Let ${}^0f_{M,R}$ be the finite part of the eigenfunction expansion of ${}^0f$ with $|m| \leq M$, $|s_j| \leq R$. Then

$$\|{}^0f - {}^0f_{M,R}\|_\infty \leq C(f) \sum_{|m|\geq M,\ |s_j|\geq R} (1 + |s_j| + |m|)^{-N} \qquad (\forall\ N).$$

(ii) With a similar notation for ${}^\theta f$,

$$\|{}^\theta f - {}^\theta f_{M,R}\|_\infty \leq C(f) \sum_{|m|\geq M} \int_{\substack{\mathrm{Re}\ s=1/2 \\ |s|\geq R}} (1 + |s| + |m|)^{-N} \left|\frac{\Delta'}{\Delta}(s)\right| ds.$$

<u>Proof</u>. (i) The key point is to have a Sobolev estimate for the sup-norm $\|u_{j,m}\|_\infty$. But some care must be taken, since $\Gamma\backslash h$ does not have a positive injectively radius (see [A], pp. 44-46).

However, it is well known that $u_{j,m}$ tend to zero in the ends of $\Gamma\backslash h$. So for any (j,m) there exists A so that $|u_{j,m}| \leq 1$ outside $F_A$. The Sobolev inequality in dimension two, $\|f\|_\infty \leq C\,\|(1-\Delta)^3 f\|_{L^2}$, applies to the part of $u_{j,m}$ smoothly cut off to $F_A$. The resulting norm at the right has the form $C(1 + |s_j|^2 + |m|^2)^3$, with constant independent of (j,m.A). The statement in (i) follows.

(ii) The argument in (i) needs to be modified since $E(\cdot,\frac{1}{2}+ir)_{(m)}$ is not even bounded. In fact, the terms $y^{\frac{1}{2}\pm ir}$ from its zero$^{th}$ Fourier coefficients (suitably cut off to the complement of a) even give counterexamples to the Sobolev inequality.

We deal with this by breaking up $E(\cdot,\frac{1}{2}+ir)_{(m)}$ into some standard pieces. First, pick A so that the complement of $F_A$ is a union of standard cuspidal ends. Choose a smooth cut off $\rho_A$ equal to 1 on $F_A$ and zero off $F_{A+1}$. Then $\rho_A({}^{\theta}f - {}^{\theta}f_{M,R})$ is supported in a compact set, so we can apply the Sobolev inequality to get

$$\|\rho_A({}^{\theta}f - {}^{\theta}f_{M,R})\|_\infty$$
$$\leq C(f) \sum_{m\geq M}\int_{\mathrm{Re}\ s=1/2} (1 + |s| + |m|)^{-N} \left|\frac{\Delta'}{\Delta}(s)\right|^{1/2} \|\rho_A E(\cdot,s)_{(m)}\| \quad (\forall\ N).$$

It is a well known consequence of the Maass-Selberg relations that $\|\rho_A E(\cdot,s)_{(m)}\| \leq C(A)\left|\frac{\Delta'}{\Delta}(s)\right|^{1/2} m$ (the m-dependence is not optimal; see [Sa], p.727 for a closely related estimate). Hence this part has the form stated in

(ii) above.

To deal with the $(1-\rho_A)$ term, we expand $(1-\rho_A)E(\cdot,s)_{(m)}$ in an invariantly defined Fourier series. This has the form $(1-\rho_A)(a_0(y,s)_{(m)} + G(z,s)_{(m)})$, where we pretend that there is just one cusp at $\infty$. The $a_0$ term contributes $(1-\rho_A)\sum_{|m|\geq M}\int_{\mathrm{Re}\ s=1/2} \hat{f}(s,m)a_0(y,s)_{(m)}\ ds$. We may view this term as a function on the (unit tangent bundle of the) Euclidean cynlinder $S^1 \times \mathbb{R}^+$, equipped with metric $d\theta^2 + (\frac{dy}{y})^2$. The injectivity radius of this is one, and the Sobolev estimate applies. Furthermore, the factors $y^{\frac{1}{2}\pm ir}$ appearing in $a_0$ form a generalized orthonormal basis for $L^2(\mathbb{R}^+,\frac{dy}{y^2})$. it follows that $\|(1-\rho_A)(^{\theta}f - {}^{\theta}f_{M,R})\|_{\infty}$ is majorized by $\sum_{|m|\geq M}\int_{\mathrm{Re}\ s=1/2} |\hat{f}(s,m)|^2(1 + |s|^2 + |m|^2)^3\ ds$. (Here we have also used unitarity of $\Phi(s)_m$). This again has the form state in (ii).

The third and final part of (ii) comes from the $(1-\rho_A)G(\cdot,s)_{(m)}$ terms. We estimate the contribution of these to $(1-\rho_A)(^{\theta}f - {}^{\theta}f_{M,R})$ by $\sum_{|m|\geq M}\int_{\mathrm{Re}\ s=1/2} |\hat{f}(s,m)|\|(1-\rho_A)G(\cdot,s)_{(m)}\|\ ds$. Here, $(1-\rho_A)G(\cdot,s)_{(m)}$ is well known to tend to zero at infinity. As in the cuspidal case, we pick B so that $|G(\cdot,s)_{(m)}| \leq 1$ on $y \geq B$. We then apply the Sobolev inequality in $F_B$. $G(\cdot,s)_{(m)}$ is an eigenfunction of the Laplacian on $y \geq A$, so we get:

$$\begin{aligned}\|(1-\rho_A)\rho_B G(\cdot,s)_{(m)}\|_{\infty} &\leq (1 + |s|^2 + |m|^2)^3\|(1-\rho_A)\rho_B G(\cdot,s)_{(m)}\|_{L^2} \\ &\leq (1 + |s|^2 + |m|^2)^3\|(1-\rho_A)G(\cdot,s)_{(m)}\|_{L^2}.\end{aligned}$$

It now follows again by the Maass-Selberg inner product formula that

$\|(1-\rho_A)G(\cdot,s)_{(m)}\|_{L^2} \leq m\left|\frac{\Delta'}{\Delta}(s)\right|^{1/2}$ (loc. cit.). As in the cuspidal case, this contributes a term of the stated form in (ii).

This completes the proof. ∎

We can now prove:

<u>Theorem 6.4</u>. Let $f \in C_0(\Gamma\backslash G)$. Then $\mu_T \to \mu(f)$ $(T \to \infty)$.

<u>Proof</u>. We may assume $f \in C_0^\infty$. But then it is clear that $\mu_T({}^0f_{M,R}) \to \mu({}^0f_{M,R})$ and $\mu_T({}^\theta f_{M,R}) \to \mu({}^\theta f_{M,R})$. The first is immediate from Theorem 6.1, while in the second we only need to add the continuity of the s-dependence of the O-symbols to that theorem.

Hence, we have only to show that $\|{}^0f - {}^0f_{M,R}\|_\infty$ and $\|{}^\theta f - {}^\theta f_{M,R}\|_\infty$ are arbitrarily small if M and R are large enough. For the cuspidal pieces this is directly visible from Prop. 6.3(i). For the θ-piece, we only need to add that $\int_0^R \left|\frac{\Delta'}{\Delta}(\frac{1}{2}+ir)\right| dr \ll R^2$ ([Sa], p. 736). ∎

Our final result is:

<u>Corollary 6.5</u>. $\mu_T \to \mu$ weakly on $C_b(\Gamma|G)$.

<u>Proof</u>. (Compare [Ito], p. 72): Let $f \in C_b$ and $\rho_A \in C_0^\infty(\Gamma\backslash G)$, $\rho_A \equiv 1$ on $F_A$, $\rho_A \equiv 0$ on the complement of $F_{A+1}$. Then:

$$|\mu_T(f) - \mu(f)| \leq \mu_T(|f - \rho_A f|) + \mu(|f - \rho_A f|) + |\mu_T(\rho_A f|) - \mu(\rho_A f)|.$$

Obviously $f - \rho_A f \in C_b$ and $\|f - \rho_A f\| \leq \|f\|_\infty$. Let $g_A \in C_b$ be equal to $(1 - \rho_A)\|f\|_\infty$. Then $|f - \rho_A f| \leq g_A$, so

$$\mu_T(|f - \rho_A f|) \leq \mu_T(g_A) = \|f\|_\infty - \mu_T(\|f\|_\infty - g_A) \longrightarrow \mu(g_A).$$

Consequently, $\limsup_{T\to\infty} \mu_T(|f - \rho_A f|) \leq \mu(g_A)$. Since $|\mu_T(\rho_A f) - \mu(\rho_A f)| \longrightarrow 0$, we have:

$$\limsup_{T\to\infty} |\mu_T(f) - \mu(f)| \leq 2\mu(g_A) \leq 2\|f\|_\infty \, \mu(F_A^c)$$

($F_A^c$ = complement of $F_A$). The right side tends to zero as $A \longrightarrow \infty$, concluding the proof.

■

## Appendix: Notations, Definitions and Background

A1. G: $PSL_2(\mathbb{R})$

$\Gamma$: cofinite subgroup

h: upper half-plane, with metric $ds^2 = y^{-2}(dx^2+dy^2)$

D: unit disc, identified with h via the Cayley map ([L], p. 85)

dg: Haar measure on G, or on $\Gamma\backslash G$

dz: area form on h, or on $\Gamma\backslash h$

$\langle, \rangle$: $L^2$-inner product on $\Gamma\backslash G$, or on $\Gamma\backslash h$

$\mathcal{F}$ (resp. F): fundamental domain for $\Gamma$ in G (resp. h)

$|F| = |\Gamma| = \text{area}(\Gamma\backslash h) = \text{vol}(\Gamma\backslash G)$

$\Omega$ (resp. $\Delta$): Casimir for G (resp. Laplacian on h)

$R = (-(\Delta+\frac{1}{4}))^{1/2}$

A2. W, H, $X_+$, V $\epsilon$ $sl_2(\mathbb{R})$: $\begin{bmatrix} 0 & 1 \\ -1 & 0 \end{bmatrix}$, $\begin{bmatrix} 1 & 0 \\ 0 & -1 \end{bmatrix}$, $\begin{bmatrix} 0 & 1 \\ 0 & 0 \end{bmatrix}$, $\begin{bmatrix} 0 & 1 \\ 1 & 0 \end{bmatrix}$

$k(\theta) = \exp \theta W$; $a_t = \exp \frac{1}{2}tH$, $a(y) = a_{\ln y}$; $n(x) = \exp xX_+$

K; A; N denote the corresponding subgroups. The same notation will be used for the left invariant vector fields (etc.) on G, or their images on $\Gamma\backslash G$. $E^{\pm}$ denote $H \pm iV$.

A3. $g \neq e$ in G is (i) elliptic, (ii) hyperbolic, (iii) parabolic if g is conjugate to an element of (i) K, (ii) A, (iii) N. For hyperbolic g, $\ell(g) = \ell$ means g is conjugate to $a_\ell$. We will also write $a(\gamma)$ for the element of A which is conjugate to a hyperbolic element $\gamma$; and $k(\gamma)$ for the element of K conjugate to an elliptic $\gamma$.

A4. $\Gamma_\gamma$ (resp. $G_\gamma$): centralizer of $\gamma$ in $\Gamma$ (resp. G)

$\gamma_0$: generator of $\Gamma_\gamma$

$\bar{\gamma}_0$: $\Gamma_\gamma \backslash G_\gamma$ (viewd as a parametrized curve in $\Gamma\backslash G$

$\int_{\bar{\gamma}_0} f$: the integral of $f \in C(\Gamma\backslash G)$ over $\bar{\gamma}_0$

Closed geodesic: $\bar{\gamma}_0$, or several times around $\bar{\gamma}_0$, or a projection of such to $\Gamma\backslash G$

$L_{\bar{\gamma}}$: length of the closed geodesic $\bar{\gamma}$, equal to $\ell(\gamma)$

Remark. The bar notation will be frequently dropped, as will the distinction between hyperbolic $\gamma$ and the associated closed geodesic, when no confusion seems possible.

A5. $\{\kappa_1, \cdots, \kappa_n\}$: complete set of inequivalent cusps for $\Gamma$

$\Gamma_a$: stabilizer of $\kappa_a$

$\sigma_a$: element of G so that $\sigma_a \cdot \infty = \kappa_a$, $\sigma_a^{-1}\Gamma_\gamma\sigma_a = \Gamma_\infty = \{ \begin{bmatrix} 1 & n \\ 0 & 1 \end{bmatrix} : n \in \mathbb{Z}\}$

$y_a(z)$: $\operatorname{Im} \sigma_A(z)$

$F = F_A \cup \bigcup_{a=1}^{h} F_A^c(a)$ where

$F_A = \{ z \in F: y_a(z) \le A, \ a = 1,\cdots,h\}$

$F_A^c(a)$ is the component of $F-F_A$ corresponding to the cusp $k_a$ (cf. [V], p. 13).

A6. $\chi: \Gamma \longrightarrow U(1)$: unitary character

$L^2_\chi(\Gamma\backslash G)$: $f(\gamma g) = \chi(\gamma)f(g)$, $f \in L^2(\mathcal{F})$

Singular cusp $\kappa_a$ for $\chi$: $\chi \equiv 1$ on $\Gamma_a$

$h(\chi)$: # singular cusps

A7. $H_n$: forms of weight n on $\Gamma\backslash G$ ($Wf = \mathrm{in}f$)

A8. $\theta_a$: $C_0^\infty(N\backslash G) \longrightarrow C_0^\infty(\Gamma\backslash G)$: $\theta$-transform ([L], p. 240; [K], p. 76)

$\theta_a^0$: adjoint of $\theta_a$ ([L], p. 241; [K], p. 84)

$\theta_a^n$: $n^{th}$ Fourier coefficient operator for $a^{th}$ cusp

$\theta_a^n[f](g) = \int_0^1 f(\sigma_a n(x)g)e(-nx)dx$, with $e(x) = e^{2\pi ix}$

$\boldsymbol{\theta}_a$: space of incomplete $\theta$-series for $a^{th}$ cusp, i.e. closure of span of $\{\theta_a(\psi)\}$

$\boldsymbol{\theta}$: $\sum_a \boldsymbol{\theta}_a$

$E_a(g,s)_{(n)}$: Eisenstein series of weight n for $a^{th}$ cusp ([K], VI)

$E_a^*(g,s_j)_{(n)}$: residue of $E_a(g,s)_{(n)}$ at a pole $s = s_j$

$M(a)$: number of poles of $E_a(g,s)$ in $(\frac{1}{2},1)$

$\{\theta_a(\cdot,s_j)\}$: orthogonal basis for the span of $\{E_a^*(\cdot,s_j)$, $a = 1,\cdots, h\}$.

A9. Cusp form $\theta_a^0(f) = 0$ for all $a$

${}^0L_\chi^2(\Gamma\backslash G)$: square integrable cusp forms

$\{u_j(\cdot,\chi)\}$: fixed orthonormalized basis of weight 0 cuspidal eigenfunctions of $\Omega$ (or $\Delta$)

$\{u_{j,m}(\cdot,\chi)\}$: weight m cuspidal eigenfunctions of $\Omega$, related to $u_j(\cdot,\chi)$ by $E^\pm u_{j,m} = (s_j \pm m + 1)u_{j,m\pm 2}$ ([L], pp. 102, 177)

$\lambda_j$, $s_j$, $r_j$, $t_j$: eigenvalues defined by $\Delta u_j = -\lambda_j u_j$, $\lambda_j = s_j(1-s_j)$, $s_j = \frac{1}{2} \pm ir_j$, $t_j = ir_j$, with $\mathrm{Im} s_j \geq 0$, $\mathrm{Re} s_j \geq 1/2$.

Complementary series, or small eigenvalues: $\lambda_j < \frac{1}{4}$

Remark. The notations $s_j$ here differs from that of [Z2]-[Z3].

A10. Fourier expansions in cusps: For each cusp $\kappa_a$, there are Fourier coefficient $\{\rho_{ja}(n)$, $\rho_{ja}(n)_{(m)}$, $\phi_{a\beta n}(s)_{(m)}\}$ so that if $g = n(x)a(y)k(\theta)$:

(i) $u_j(\sigma_a z) = y^{1/2} \sum_{n\neq 0} \rho_{ja}(n)K_{it_j}(2\pi|n|y)e(nx)$ ([Ku], (3.38))

(ii) $u_{j,m}(\sigma_\alpha g) = e^{im\theta} \sum_{n\neq 0} \rho_{j\alpha}(n)_{(m)} W_{s_j,m}(2\pi|n|y)e(nx)$ ([F], (66))

(iii) $E_\alpha(\sigma_\beta z,s) = a_{0\alpha\beta}(y,s) + G_{\alpha\beta}(\sigma_\beta z,s)$

$a_{0\alpha\beta}(y,s) = \delta_{\alpha\beta}y^s + \phi_{\alpha\beta}(s)y^{1-s}$

$G_{\alpha\beta}(\sigma_\beta z,s) = y^{1/2} \sum_{n\neq 0} \phi_{\alpha\beta n}(s)\, K_{s-(1/2)}(2\pi|n|y)e(nx)$ ([K], p. 45)

(iv) $E_\alpha(\sigma_\beta g,s)_{(m)} = a_{0\alpha\beta}(g,s)_{(m)} + G_{\alpha\beta}(\sigma_\beta g,s)_{(m)}$

$a_{0\alpha\beta}(g,s)_{(m)} = (\delta_{\alpha\beta}y^s + \phi_{\alpha\beta}(s)_{(m)}y^{1-s})\, e^{im\theta}$

$G_{\alpha\beta}(\sigma_\beta g,s)_{(m)} = e^{im\theta} \sum_{n\neq 0} \phi_{\alpha\beta n}(s)\, y^{1-s}\, W_{s,m}(2\pi|n|y)$

$W_{s,m}(y) = \int_{\infty}^{\infty} (\frac{t+i}{|t+i|})^m (1+t^2)^{-s} e(-ty)dt$ (cf. [K], pp. 65-66; [F], § 3)

(v) $\Phi(s)_{(m)}$: $(\phi_{\alpha\beta}(s)_{(m)})$; $\Delta(s)_{(m)} = \det \Phi(s)_m$ .

A11. $\psi_m$: lowest weight vector from a holomorphic discrete series irreducible $\mathcal{D}_m^+$ in $L^2_\chi(\Gamma\backslash G)$

A12. $Op(\sigma)$: $\psi$DO (pseudo-differential operator) with complete symbol $\sigma$ (cf. [Z2-3])

$\langle Op(\sigma)u_j, u_i\rangle$: matrix element $Op(\sigma)$ relative to $\{u_j\}$; similarly for $\langle Op(\sigma)E(\cdot,\frac{1}{2}+ir), E(\cdot,\frac{1}{2}+ir)\rangle$

A13. Rankin-Selberg zeta functions (cf. [Za2]):

(a) $R_\alpha[|u_j|^2, s]$: $\langle Op(E_\alpha(\cdot,s))u_j,u_j\rangle = \int_F E_\alpha(z,s)|u_j(z)|^2dz =$

$\frac{\Gamma^2(\frac{s}{2})\Gamma(\frac{s}{2}+it_j)\Gamma(\frac{s}{2}-it_j)}{4\pi^s} R_{j,\alpha}(s)$, with $R_{j,\alpha}(s) = \sum_{n\neq 0} |\rho_{j\alpha}(n)|^2|n|^{-s}$

(Re s > 1)

(b) $R_\alpha[|u_j|^2, s]_m = \langle Op(E_\alpha(\cdot,s))_{(m)})u_j,u_j\rangle = \Gamma_j^m(s)^{-1}R^m_{j,\alpha}(s)$, with

$R_{j,a}(s) = \sum_{a\neq 0} (\rho_{ja}(n)\overline{\rho_{ja}(n)_{(m)}})n^{-s}$ (Re s > 1) and with

$(\Gamma_j^m(s))^{-1} = \int_0^\infty y^{s-2} W_{j,0}(2\pi y) W_{j,m}(2\pi y)dy$. Formula (b) follows by a standard unfolding argument: $\langle Op(E_a(\cdot,s))_{(m)})u_j,u_j\rangle = \int_{\mathcal{F}} E_a(g,s)_{(m)}u_j(g)\bar{u}_{j,m}(g)dg = \int_{\mathcal{F}_a} y(g,s)_{(m)}u_j(g)\bar{u}_{j,m}(g)dg$ ($\mathcal{F}_a$ is a fundamental domain for $\Gamma_a$ in G; for $y(g,s)_{(m)}$ see ([K], (6.1.1)). Unfolding, the latter is

$\int_0^\infty y^{s-2}[\sum_{n\neq 0} \rho_{ja}(n)\overline{\rho_{ja}(n)_{(m)}}\, W_{j,0}(2\pi|n|y)W_{j,m}(2\pi|n|y)dy =$
$(\Gamma_j^m(s))^{-1}R_{j,a}^m(s)$.

(c) The analogues for Eisenstein series are:

$$R_{a\beta}(\tau,s) = \sum_{n\neq 0} \frac{|\phi_{a\beta n}(\tau)|^2}{|n|^s} \quad (\text{Re } s > 1).$$

Then

$$R_{a\beta}(\tfrac{1}{2}+it,s) = \frac{4\pi^s}{\Gamma(\frac{s}{2})^2\Gamma(\frac{s}{2}+it)\Gamma(\frac{s}{2}-it)}\, R_\beta(|E_a(\cdot,\tfrac{1}{2}+it)|^2,s),$$

where

$$R_\beta(|E_a(\cdot,\tfrac{1}{2}+it)|^2,s) = \int_0^\infty y^{s-2}\theta_\beta^0[|G_{a\beta}(\cdot,\tfrac{1}{2}+it)|^2](y)dy$$

(d) For residues of Eisenstein series, we set:

$$R^*_{a,\beta\gamma}(s_j,s,\chi) = \sum_{n\neq 0} \frac{\phi^*_{\beta a n}(s_j,\chi)\,\overline{\phi^*_{\gamma a n}(s_j,\chi)}}{|n|^s} \quad (\text{Re } s > 1)$$
$$= \frac{4\pi^s}{\Gamma(\frac{s}{2})^2\Gamma(\frac{s}{2}+(s_j-\frac{1}{2}))\Gamma(\frac{s}{2}-(s_j-\frac{1}{2}))}\, R_a[E^*_\beta(\cdot,s_j)E^*_\gamma(\cdot,s_j),s],$$

with

$$R_a[E^*_\beta(\cdot,s_j)E^*_\gamma(\cdot,s_j),s] = \int_0^\infty y^{s-2}\theta_a^0[G^*_{\beta a}(\cdot,s_j)\overline{G}^*_{\gamma a}(\cdot,s_j)](y)dy$$

A14. Convolution operators

$R_g$ (resp. $R_g^\Gamma$): right translation on $L^2(G)$ (resp. $L^2(\Gamma\backslash G)$)

$R_\phi$ (resp. $R_\phi^\Gamma$): $\int_G \phi(g) R_g dg$ (resp. $R_g^\Gamma$)

$S_{m,0}$: $\{\phi \in C(G): \phi(k(\theta_1)gk(\theta_2)) = e^{im\theta_1}\phi(g)\}$

Note. $\phi \in S_{0,0}$ will often be identified with a corresponding function (still denoted $\phi$) of the real variable tr $g^t g-2$ ([Za1], [L], p. 72). The kernel of $R_\phi^\Gamma$ can be then identified with a point pair invariant kernel $k(z,z') = \phi[\frac{|z-z'|^2}{yy'}]$ on $h \times h$ ([Za1]). If $\Delta u = -(\frac{1}{4}+r^2)u$, then $R_\phi^\Gamma u = h(r)u$, where $h$ is the Selberg transform of $\phi$ ([Za1], p. 319). Let $R = \sqrt{-[\Delta+\frac{1}{4}]}$; we then write $R_\phi^\Gamma = h(R)$ in the above case.

A15. Spherical transforms

$\Phi_{m,s}$: generalized spherical function in $S_{m,0}$ (denoted $\Phi_{\lambda,-(m/2)}$, with $s = i\lambda$, in ([He], p. 50 (46); compare [Z1], (1.9.2)).

$S_m$: weight $m$ spherical transform:

$S_m\phi(s) = \int_G \phi(g)\Phi_{-m,s}(g)dg$ $(\phi \in C_0^\infty \cap S_{m,0})$

$S_m^{-1}$: inverse spherical transform:

$S_m^{-1}g(x) = \int_{\mathrm{Re}\ s=0} g(s)\Phi_{m,s}(x)\ (\frac{s}{i}\tanh(\frac{\pi s}{i}))\ \frac{ds}{2\pi i}$

Background: $S_m$: $C_0^\infty \cap S_{m,0} \longrightarrow PW_m$ is a bijection, where $PW_m$ is the Paley-Wiener space of entire $f$ satisfying: $|f(\sigma+it)| << e^{k|\sigma|}(1+|t|)^{-N}$ for some $k$ and all $N$; and also: $f(-s) = \frac{P_m(s)}{P_m(-s)} f(s)$, where $P_m(s) = (|m|-1+\frac{1}{2}(s+1))\cdots(\frac{1}{2}(s+1))$. See [He, Theorem 4.2].

A16. Harish-Chandra (HC) transforms

(a) $H_m$: weight $m$ HC transform: $C_0^\infty \cap S_{m,0} \longrightarrow C_0^\infty(A)$,

$H_m f(a) = |D(a)| \int_{A\backslash G} f(x^{-1}ax)\chi_m(x)dx$, where $D\begin{bmatrix} a & 0 \\ 0 & a^{-1} \end{bmatrix} = a - a^{-1}$

and $\chi_m(a_n k(\theta)) = e^{im\theta}$.

$\mathbf{M}$ = Mellin transform: $C_0^\infty(A) \longrightarrow PW(\mathbb{C})$ ([L], V. 3).

<u>Background</u>. $S_m = \mathbf{M}H_m$ ([Z2], (1.14))

(b) $\{H_{m,s};\ H_k^c,\ H_{k,+}^c,\ H_m^d;\ L_k^c\}$: non-standard conjugacy class transforms arising in the generalized trace formulae of [Z2] and of this paper. The H-transforms involve hyperbolic conjugacy classes as in (a), while L is an ellptic class transform.

The hyperbolic transforms are all offspring of the transform $H_{m,s}$:

$(b.i)_{(m,s)}$ $\quad H_{m,s}: C_0^\infty \cap S_{m,0} \longrightarrow C_0^\infty(A)$,

$$H_{m,s}f(a) = |D(a)| \int_{A\backslash G} f(x^{-1}ax)\chi_m(x)\mathcal{F}_s(x)dx,$$

where $\mathcal{F}_s$ is a certain special (hypergeometric) function which we will not need to specify here.

The transforms which actually appear in the final versions of the trace formulae (Theorems 2(A)-(C) and 3(A)-(B); [Z-2], Propositions 2.10-2.12) are certain modifications $\{H_k^c,\ H_{k,+}^c,\ H_m^d\}$ of $H_{m,s}$.

First, $H_k^c$ is a modification of $H_{0,s_k}$, where $\{s_k\}$ parametrize the continuous series representations in $L^2(\Gamma\backslash G)$ (cf. § 1.9):

$(bii)_{(0,s_k)}$ $\quad H_k^c: C_0^\infty \cap S_{0,0} \longrightarrow C_0^\infty(A)$,

$$H_k^c\,\varphi(a) = \int_{-\infty}^{\infty} \varphi(n(x)^{-1}an(x)F_{s_k}(-x^2)dx,$$

where

(biii) $$F_{s_k}(u) = F(\tfrac{1}{4} + \tfrac{s_k}{4}, \tfrac{1}{4} - \tfrac{s_k}{4}, \tfrac{1}{2}, u)$$

(the standard hypergeometric function).

As mentioned in item 14, $\phi$ may be identified with a function on $\mathbb{R}^+$, and $H_k^c$ may then be viewed as a transform on $C_0^\infty(\mathbb{R}^+)$. This is described in detail in [Z-2, § 3].

Second, the transform $H_{k,+}^c$ are modifications of a kind of imaginary part of $H_{2,s_k}$ (the + stands for $X_+$; see [Z-2], § 2 for details on the imaginary part). Viewed as transforms on $C_0^\infty(\mathbb{R}^+)$, they are defined by:

(biv)$_{(+,s_k)}$ $$H_{k,+}^c\colon C_0^\infty(\mathbb{R}^+) \longrightarrow C_0^\infty(\mathbb{R}^+),$$

$$H_{k,+}^c\, \varphi(v) = (v+2)^{1/2} \int_{-\infty}^{\infty} \varphi(v(u^2+1) + 2)(v(u^2+1)+2)^{1/2} \cdot (u^2+1) F_{k,+}(-u^2)\, du,$$

where

(bv)$_{(+,s_k)}$ $$F_{k,+}(x) = F(\tfrac{3}{4} + \tfrac{s_k}{4}, \tfrac{3}{4} - \tfrac{s_k}{4}, \tfrac{1}{2}, x),$$

(see [Z-2], Proposition 2.11).

Third, the transform $H_m^d$ is a modification of $H_{m,m}$ corresponding to the discrete series parameter m. Viewed as a transform on $C_0^\infty(\mathbb{R})$, it is defined by:

(bvi)$_m$ $$H_m^d\, \varphi(v) = v^{\frac{m}{4} - \frac{1}{2}} \int_{-\infty}^{\infty} \varphi(1 + \frac{v-1}{u^2+1})\ (1 + \frac{v-1}{u^2+1})^{-m/4}\ (u+i)^{-m/2} du$$

(see [Z.2], Proposition 2.12).

Finally, the elliptic HC transforms $L_k^c\colon C_0^\infty \cap S_{0,0} \longrightarrow C^\infty(K)$ are defined by:

$$(\mathrm{bvii})_k \qquad L_k\varphi(K(\theta)) = \int_A \varphi(a^{-1}k(\theta)a)\ \varphi_{s_k}(a)\ d\mu(a)$$

where $d\mu(a) = \frac{1}{2} D(a)da$, and where $\varphi_{s_k}$ is the spherical function (cf. [He]).

A17. Spectral decomposition

(a) $L^2(\Gamma\backslash G) = {}^0L^2 \oplus \theta$ ([L], XII-XIII)

(b) $\theta = L^2_{eis} \oplus L^2_{res} \oplus \mathbb{C}$ (span of Eisenstein series, resp. residues, resp. constants)

(c) $\pi_0$, $\pi_{eis}$, $\pi_{res}$, $\pi_1$: corresponding orthogonal projections on ${}^0L^2$, $L^2_{eis}$, $L^2_{res}$, $\mathbb{C}$ (resp.)

## References


[A] T. Aubin, Nonlinear Analysis on Manifolds, Monge-Ampère Equations, Springer-Verlag (1980).

[B] R. Bowen, The equidistribution theory of closed geodesics, Am. J. Math. 94 (1972), 413-423.

[B-K] K. Burns and A. Katok, Manifolds of non-positive curvature, Ergod. Th. of Dynam. Sys. 5(1985), 307-317.

[C-S] P. Cohen and P. Sarnak, Mimeographed notes on the Selberg trace formula.

[DeG] D. L. De George, Ecole Normale Superieure, Annales Sc. Ser. 4, 10(1977), p. 133

[D-I] J.-M. Deshouillers and H. Iwaniec, The non-vanishing of Rankin-Selberg zeta-functions at special points, Contemporary Math. (AMS), vol. 53, 1986.

[E] I. Efrat, Cusp forms in higher rank, to appear.

[F] J. Fay, Fourier coefficients of the resolvent for a Fuchsian group, J. fur. die Reine und Augew, Math. 293, 143-203 (1977).

[G-R] I. S. Gradshteyn and I. M. Ryzik, Tables of Integrals, Series and Products, 4th ed., Academic Press, 1980.

[G-W] R. Gangolli and G. Warner, Zeta functions of Selberg's type for some non-compact quotients of symmetric spaces of rank one, Nagoya Math. J. 78(1980), 1-44.

[Hej] D. Hejhal, The Selberg Trace Formula for $PSL_2(\mathbb{R})$, vol. 2, SLN 1001, Springer Verlag, Berlin (1983).

[He] S. Helgason, Groups and Geometric Analysis, Academic Press (1984).

[Ito] K. Ito, Introduction to Probability Theory, Cambridge Univ. Press (1984).

[Iv] A. Ivič, The Riemann Zeta-Function, John Wiley & Sons (1985).

[I.1] H. Iwaniec, Prime geodesic theorem, J. Reine Angew. Math. 349 (1984), 136-159.

[K] T. Kubota, Elementary Theory of Eisenstein Series, Kodansha, Ltd., Tokyo and John Wiley & Sons, New York (1973).

[Ku] N. V. Kuznecov, Petersson's conjecture for cusp forms of weight 0 and Linnik's conjecture. Sums of Kloosterman sums, Math. USSR Sbornik

39:3 (1981).

[L] S. Lang, $SL_2(\mathbb{R})$, Addison-Wesley (1975).

[P] W. Parry, Bowen's equidistribution theory and the Dirichlet density theory, Ergod. Th. of Dynam. Sys. 4(1984), 117-134.

[P-S] R. Phillips and P. Sarnak, On cusp forms for co-finite subgroups of $PSL_2(\mathbb{R})$, Invent. Math. 80 (1984), 339-364.

[Po] M. Pollicott, Meromorphic extensions of generalized zeta functions, Inv. Math. 85(1986), 147-164.

[Ro] W. Roelcke, Das Eigenwertproblem der automorphen Formen in der hyperbolischen Ebene, II, Math. Ann. 168 (1967), 261-324.

[Sa.1] P. Sarnak, The arithmetic and geometry of some hyperbolic three manifolds, Acta. Math. (1982), 252-295.

[Sa.2] P. Sarnak, Horocycle flow and Eisenstein series, Comm. Pure Appl. Math. 34 (1981), 719-739.

[Su] D. Sullivan, Discrete conformal groups and measurable dynamics, BAMS 6:1 (1982), 57-75.

[V] A. B. Venkov, Spectral Theory of Automorphic Functions, Proc. Steklov Inst. Math., Issue 4 (1982).

[Za1] D. Zagier, Eisenstein series and the Selberg Trace Formula I, in Automorphic Forms, Representation Theory and Arithmetic, Bombay (1979), Springer-verlag, Berlin-Heidelberg-New York (1981), pp. 305-355.

[Za2] D. Zagier, The rankin-Selberg method for automorphic functions which are not of rapid decay, J. Fac. Sci., Univ. Tokyo, Sect. 1A, vol. 28 (1981), 415-439.

[Z1] S. Zelditch, Trace formulae for compact $\Gamma\backslash PSL_2(\mathbb{R})$ and the equidistribution theorems for closed geodesics and Laplace eigenfunctions, Springer Lecture Notes (vol. 1256).

[Z2] ———, Trace formulae for compact $\Gamma\backslash PSL_2(\mathbb{R})$ and the equidistribution theory of closed geodesics, Duke Math. J. vol. 59(1989), 27-81.

[Z3] ———, Pseudo-differential operators, trace formulae and the geodesics integrals of automorphic forms, Duke Math. J., vol. 56 (1988), 295-344.

[Z4] ———, Uniform distribution of eigenfunctions on compact hyperbolic surfaces, Duke Math. J., vol. 55 (1987), 919-941.

[Z5] ———————, Mean Lindelöf hypothesis and equidistribution theorems for cusp forms and Eisenstein series (J. Fun. Anal., to appear).

[Z6] ———————, Pseudo-differential analysis on hyperbolic surfaces, J. Fun. Anal. vol. 68(1986), 72-105.

## MEMOIRS of the American Mathematical Society

**SUBMISSION.** This journal is designed particularly for long research papers (and groups of cognate papers) in pure and applied mathematics. The papers, in general, are longer than those in the TRANSACTIONS of the American Mathematical Society, with which it shares an editorial committee. Mathematical papers intended for publication in the Memoirs should be addressed to one of the editors:

**General instructions to authors for**

## PREPARING REPRODUCTION COPY FOR MEMOIRS

**For more detailed instructions send for AMS booklet, "A Guide for Authors of Memoirs." Write to Editorial Offices, American Mathematical Society, P.O. Box 6248, Providence, R.I. 02940-6248.**

**MEMOIRS** are printed by photo-offset from camera copy fully prepared by the author. This means that the finished book will look exactly like the copy submitted. Thus the author will want to use a good quality typewriter with a new, medium-inked black ribbon, and submit clean copy on the appropriate model paper.

**Model Paper**, provided at no cost by the AMS, is paper marked with blue lines that confine the copy to the appropriate size.

**Special Characters** may be filled in carefully freehand, using dense black ink, or **INSTANT** ("rub-on") **LETTERING** may be used. These may be available at a local art supply store.

**Diagrams** may be drawn in black ink either directly on the model sheet, or on a separate sheet and pasted with rubber cement into spaces left for them in the text. Ballpoint pen is not acceptable.

**Page Headings** (Running Heads) should be centered, in CAPITAL LETTERS (preferably), at the top of the page — just above the blue line and touching it.

LEFT-hand, EVEN-numbered pages should be headed with the AUTHOR'S NAME;

RIGHT-hand, ODD-numbered pages should be headed with the TITLE of the paper (in shortened form if necessary).

Exceptions: PAGE 1 and any other page that carries a display title require NO RUNNING HEADS.

**Page Numbers** should be at the top of the page, on the same line with the running heads.

LEFT-hand, EVEN numbers — flush with left margin;

RIGHT-hand, ODD numbers — flush with right margin.

Exceptions: PAGE 1 and any other page that carries a display title should have page number, centered below the text, on blue line provided.

FRONT MATTER PAGES should be numbered with Roman numerals (lower case), positioned below text in same manner as described above.

## MEMOIRS FORMAT

**It is suggested that the material be arranged in pages as indicated below.
Note: Starred items (*) are requirements of publication.**

**Front Matter** (first pages in book, preceding main body of text).

Page i — *Title, *Author's name.

Page iii — Table of contents.

Page iv — *Abstract (at least 1 sentence and at most 300 words).

Key words and phrases, if desired. (A list which covers the content of the paper adequately enough to be useful for an information retrieval system.)

**1991 Mathematics Subject Classification*. This classification represents the primary and secondary subjects of the paper, and the scheme can be found in Annual Subject Indexes of MATHEMATICAL REVIEWS beginnning in 1990.

Page 1 — Preface, introduction, or any other matter not belonging in body of text.

Footnotes: *Received by the editor date.
Support information — grants, credits, etc.

First Page Following Introduction - Chapter Title (dropped 1 inch from top line, and centered). Beginning of Text.

Last Page (at bottom) - Author's affiliation.

# Recent Titles in This Series

*(Continued from the front of this publication)*



(See the AMS catalogue for earlier titles)

FORSYTH LIBRARY
FORT HAYS STATE UNIVERSITY